# Warriors of the Cloisters

# Warriors of the Cloisters

## The Central Asian Origins of Science in the Medieval World

Christopher I. Beckwith

PRINCETON UNIVERSITY PRESS
Princeton and Oxford

Published by Princeton University Press, 41 William Street, Princeton, New Jersey 08540

In the United Kingdom: Princeton University Press, 6 Oxford Street, Woodstock, Oxfordshire OX20 1TW

press.princeton.edu

Jacket illustration: Vittore Carpaccio (c. 1460/5–1523/6), *Debate of St. Stephen* (oil on canvas), Pinacoteca di Brera, Milan, Italy / Alinari / The Bridgeman Art Library



Library of Congress Cataloging-in-Publication Data

Beckwith, Christopher I., 1945–
Warriors of the cloisters : the Central Asian origins of science in the medieval world / Christopher I. Beckwith.
p. cm.
Includes bibliographical references and index.
ISBN 978-0-691-15531-9 (hardback : alk. paper) 1. Science, Medieval. 2. Academic disputations. 3. Science—Asia, Central—History—To 1500. I. Title.
Q124.97.B43 2012
509.58'0902—dc23
2012003721

British Library Cataloging-in-Publication Data is available

This book has been composed in Minion Pro

Printed on acid-free paper. ∞

Printed in the United States of America

1 3 5 7 9 10 8 6 4 2

**Truth is born in argument.**

—Russian saying attributed to Socrates (d. 399 bc)

# CONTENTS

# PREFACE

THIS BOOK IS ABOUT the medieval "scientific method"—the recursive argument method—and how it was transmitted along with the college from Buddhists to Muslims in Central Asia, and from there to medieval Western Europe.

Two decades ago, I first attempted to work on the history of the recursive argument method (traditionally known as the 'scholastic method', or more precisely, the 'disputed questions' method) and in 1990 published a paper on it. However, though I felt sure that there was a connection among the Latin, Arabic, and Tibetan varieties of the method, I could not explain the differences among the versions I had found and was unable to resolve the many problems connected with its origin and spread. So I abandoned the topic.

In 2007, while preparing the final manuscript of my book *Empires of the Silk Road* for publication, it occurred to me once again to look into the *Mahāvibhāṣa*, a work obtained by the great traveler-monk Hsüan Tsang when he was in the Central Asian city of Balkh in 628 or 630 and later translated by him into Chinese, in which language alone it is now preserved. I had looked at the text in or around 1990, but did not understand its metalanguage well enough to discover the argument structure used in it. This time, with the inestimable help of a translation of one section of the text done by Takeda Hiromichi and Collett Cox, I found what I had previously suspected might be there. I then reexamined the earliest Latin works that use the method, and at the same time looked into early examples of similar works in Arabic. Partly as a result of doing all this at around the same time, I realized that earlier analyses of the structure of the method, including my own, were misleading or wrong, and the direct connection of the different versions of it had therefore been overlooked. I presented the results of my discoveries in lectures given in 2008 in several locations in the United States and Europe. Between the Paris and Oxford lectures, it occurred to me that I had also missed the most characteristic and essential feature of the method: formal recursion.

Subsequently I continued to work on the topic, producing a brief article on some points as I then understood them and a rough draft of a book. There the matter remained for awhile as my duties as a teacher took up most of my time and energy for the rest of that year and the beginning of the following year. Then, though I was busy with teaching while in Vienna in the late spring and early summer of 2009, I nevertheless managed to write the first actual manuscript of the book, and to revise the above-mentioned brief article. When I was once again free to work on the project full time, in the last two months of that summer, I completely reorganized and rewrote my manuscript, and revised it that fall semester. I then submitted it to Princeton University Press in December.

In the process of further revising the manuscript, I have discovered that my analysis of the data as of the end of 2009 continues, in all significant aspects, to be confirmed, and has actually been strengthened, by everything additional I have found since. In particular, intensive examination of the putative exceptions suggested by colleagues has been interesting, and in some cases relevant, but on the whole it has ended up not having any significant effect on the book's argument. In addition, although I had always believed that the late George Makdisi was basically on the right track with regard to the history and transmission of the college and the recursive argument method (the scholastic method) from the Islamic world to medieval Western Europe, I found that some of the details of his arguments—especially those involving the education system—also seem to be confirmed, with one significant lingering exception.

The exception appears to be law, which Makdisi was convinced was the source of the Classical Arabic recursive argument method, since the Islamic colleges were devoted mostly to law. My investigations suggest it is indeed possible that there may be material relevant to the history of the recursive argument method in Medieval Latin, Classical Arabic, and early Indic texts (or their Chinese translations) concerned directly or indirectly with law. Most significantly of all, it is expressly stated by Avicenna that he learned the recursive method from a teacher of *fiqh* (Islamic jurisprudence), thus supporting Makdisi's theory. The arguments I myself have seen in my forays into the legal literature appear to be different, but law is not a subject I am trained in, familiar with, or understand very well, even in English, not to speak of any other language. Scholars specializing in *fiqh* should test Makdisi's theory by examining early "disputed questions" texts in Arabic to determine the relationship, if any, of their argument structure to that of the recursive argument method described in this book. Until that is done, Makdisi's thesis about a possible legal connection remains untested.

I like to think that I have now answered most of the major problems I set out to solve, but much remains to be done. There is certainly a great need for further study of the recursive argument method in Classical Arabic philosophical and theological literature. In addition, we need a major study of the *Aṣṭagrantha*, much more work on the *Mahāvibhāṣa* (a gigantic book), and detailed examination of Tibetan literature for possible examples of Tibetan authors' use of the Indian form of the recursive argument method. When this book was about to go into production it was suggested to me that the method had actually been adopted by early Chinese Buddhist scholars and is used in some important commentaries in that tradition, though it later fell out of use. Unfortunately, I learned this too late to be able to get a legible copy of the one study that seems to discuss this and give examples of it, so as to incorporate it into the present book. I hope that other scholars will pursue these matters. Although in several important cases the treatment of the topics covered in the present book apparently constitutes the first published attempt at analyzing them, it is not (I hope) the last word on anything. Further study is still very much needed by specialists in each relevant field—for example, early Central Asian Buddhist texts; Indian Buddhist texts; Arabic philosophical, scientific, and theological texts; Medieval Latin philosophical, scientific, and theological texts; and early Chinese Buddhist commentaries—including careful, detailed treatment and presentation of more examples of the recursive argument method. Among the most serious lacunae are an English translation of Abelard's *Sic et non*, a translation and study of Avicenna's *De anima*, and a translation of Vasubandhu's *Abhidharmakośabhāṣya* from the Sanskrit.

This book treats aspects of the history of science in Asia and in Western Europe. For medieval Europe, unlike the other regions, there is a considerable body of scholarly literature on this topic. I do not attempt to revise the current general consensus of specialists in medieval science or postmedieval science in any significant way. Instead, I mainly present new research and new analyses regarding little-studied or even untouched topics within that field, and would like to suggest that this material does have major implications for the current consensus. Therefore, although I have relied on the work of my predecessors, this book aims to present what I have been able to contribute myself, within my own limitations, and it addresses topics of interest to me and about which I have been able to learn a little, from my perspective.

In connection with the above, I would like to respond to a comment by one of the anonymous referees of the manuscript of this book. As far as I have been able to determine after much personal searching and much inquiring of scholars specializing in relevant fields, the existing scholarly literature on the specific topics to which this book is mainly devoted is extremely limited. No one has published anything, beyond the occasional short remark, on the recursive argument method of early Central Asian Buddhist literature (which is preserved mainly in Chinese translations), even when those texts have been translated into a Western language. No one has published anything on the same method in medieval Classical Arabic philosophical texts, though it has been referred to (evidently incorrectly) by a few scholars, above all by George Makdisi, apparently the only scholar to do more than mention the existence of the scholastic method in Classical Arabic. No one has noted that the recursive argument method of Medieval Latin texts (the fully developed scholastic method used by, for example, Albert the Great and Thomas Aquinas) first appears not in works originally written in Latin, but in the mid-twelfth-century Latin translations of works originally written in Arabic by the great Central Asian natural philosopher, Avicenna. And as far as I know no one has published a detailed analysis of the recursive structure of the method, the differences between it and other formats used in medieval scholastic literature in those traditions and in Medieval Latin, and so on. It would perhaps have made my task easier if there were at least a few such works, or if I had been able to discover those I may have missed. It would certainly have been easier if someone had identified which Buddhist works, Classical Arabic works, or even Medieval Latin works contain the method. But with a very few exceptions that is not the case. As is normal scholarly practice, I cite the relevant works published on the topic, including my own, but for many topics there are only source citations and my own arguments based on my analysis of the sources.

To put it another way, this book is not a bibliographical summary of what has been done before, with little new but my English style. If it were, it would be drastically shorter than it already is, considering the extremely limited secondary literature available on the topic. It would also be written by someone else, not by me, because I have no interest whatsoever in compiling anything of that sort. Most of the references that are included in the book are there strictly to help the interested reader find the sources and closely relevant scholarly literature I have used or cited, not to buttress every statement of common-knowledge historical fact.

This book is also not about ancient science, medieval scholastic thought, ancient to contemporary debates about the validity of science, or many other things that one could write a book about. Much could and should be written about such topics, and others of contemporary relevance, but there is already enough "new knowledge"—data, analyses, theories—in this book to occupy the time and energy of the self-appointed gatekeepers of several academic fields. So I have done my best to eliminate the slightest hint of such unsavory issues, and instead encourage those interested in them to read the "modern" chapters of my 2009 book and apply the ideas and analyses in them to the contemporary situation with respect to science and related subjects.

I apologize to readers who would prefer a work such as one of those described hypothetically above, but as it happens, introductory books on these topics, including some textbooks, already exist for Western European medieval studies (the only subfield in which such a problem might come up to begin with); some of them are cited here and there in the text for the benefit of interested readers.

A related observation comes to mind here. A certain amount of repetition has ended up being unavoidable. (If I were very clever I would somehow have managed to make it recursive.) This seems to be due in part to the episodic presentation both by chapters and within individual chapters, and in part to my desire not to lose the reader amidst all the detail.

A problem particular to the topic of this book is that almost everything that has been published (again nearly all of it on Medieval Latin topics) that ostensibly treats the recursive argument method—what has usually been called the "scholastic method" by earlier specialists—has little or nothing to do with that method itself. When the scholarly literature does refer to a text's argument method, it rarely distinguishes clearly, if at all, between the early *sententiae* 'sentences' or *quaestiones* 'questions' structure still used for native Latin compositions in the twelfth century, and the *quaestiones disputatae* 'disputed questions' structure (the "recursive argument method" structure), which first appears in Latin translations from Arabic in the mid-twelfth century and is first definitely used in new compositions by authors writing in Latin shortly after 1200 AD. It is certainly true that some medievalists have not overlooked this difference, but although they themselves seem to be perfectly clear on the issue, other scholars are far from clear about it. This seems mainly to be because studies on the Medieval Latin "scholastic method" can actually be about almost anything imaginable, even though they apparently specify their topic as the *quaestiones disputatae* 'disputed questions' of famous writers such as Albertus Magnus (Albert the Great), Thomas Aquinas, Roger Bacon, and many others, which are obvious, classic examples of the recursive argument method. Some works display a stunning unawareness of the differences among the radically unlike literary formats used from antiquity through the Middle Ages, all of which are often lumped together as examples of the "scholastic method." The widespread confusion seems to have a simple explanation: the *term* 'scholastic method' has become so diffuse and imprecise that it no longer conveys any useful information. I have therefore come to the realization that it is imperative to abandon it. In its place I have adopted a precise, purely descriptive term, *recursive argument method* (or 'recursive argument' or 'recursive method', for short) and have used it throughout.

The structure of the recursive argument method is recognized explicitly by several leading scholars, some of whom even provide a thumbnail outline of the method's constituent parts (see chapter 2), though they do not, of course, use my new terminology, and the recent works by Edward Grant do contain much that is illuminating on the method in connection with its use in scientific works. However, I have looked in vain for a study that analyzes *in depth* the structure of the method and what is distinctive about it, whatever an author might have thought that to be. There are some fine studies of medieval manuscripts, of related historical issues, and of the niceties differentiating various functions, ceremonial uses, or other aspects of the fully developed Medieval Latin method, but as far as I could discover there are no studies directly relevant to this or the other topics on which I have mainly focused.

Finally, it is well known that medieval science was on the whole not done by experiments in laboratories. Where or how was it done, then? The scholarly consensus among specialists is that medieval scientific activity was done in public disputations, both oral and written. These disputations used the medieval recursive argument method, the topic of the present book. Accordingly, that method was the actual "scientific method" of the Middle Ages and Renaissance. I have not attempted to trace the fate of the recursive argument method in later times, other than to show that it did not entirely disappear. It is still present in the modern concept of the ideal scientific method, and is still used practically, to some extent, in scientific research reports, which are briefly discussed in chapter 8.

I know that as always many problems still remain, and I have undoubtedly made missteps along the way. I sincerely hope that other scholars will follow up on my attempts to lay the groundwork for further research on the topics covered, and will improve on what I present here. I wish success to all who continue the quest.

# ACKNOWLEDGMENTS

MANY PEOPLE HAVE HELPED ME with different aspects of the writing and publication of this book over the past few years.

I am especially indebted to Collett Cox for kindly sending me the manuscript of her article on the *Mahāvibhāṣa* jointly authored with Takeda Hiromichi, and to Helmut Krasser for his help with the Tibetan version of the recursive argument method. I would also like to thank Pascale Hugon and Anne MacDonald for their generous and kind help with Indo-Tibetan Buddhist scholastic texts, and Georgios Halkias and S. Frederick Starr for their helpful corrections and suggestions with respect to an early draft of the manuscript.

In addition, I am grateful to Heather Stoddard, and to Fernand Meyer, my host at the École Pratique des Hautes Études (Sorbonne), for inviting me to give a long lecture on the topic on June 5, 2008. I would also like to thank Robert Mayer, who invited me to give a shorter lecture from a different perspective in the Numata Distinguished Guest Speaker Series at the Oriental Institute of Oxford University's Faculty of Oriental Studies, in association with the Oxford Centre for Buddhist Studies, on June 12, 2008. The enthusiasm, friendliness, and encouragement of those in Paris and Oxford during and after my lectures there was a great support to me in the writing of this book.

To many others who have helped me in various ways I owe my thanks as well, in particular to Rob Tempio, my editor at Princeton University Press, who strongly supported me from the moment I told him about the book, and to Salman Al-Ani, Kalani Craig, Patricia Crone, Jennifer Dubeansky, Noretta Koertge, Nicholas Kontovas, Richard Nance, Emanuel J. Mickel, Natalia Murataeva, Karin Preisendanz, Raphael Sealey, Suzanne Stetkevych, Kevin Van Bladel, Michael L. Walter, Eduard Vilella, Rega Wood, the librarians of the Wells Library at Indiana University, and the staff of the Käte Hamburger Kolleg at the Ruhr-Universität Bochum.

Most of all, I am as always grateful to my wife, Inna, who unflaggingly encouraged me in my work on the book, from beginning to end.

I would also like to thank everyone else who has helped me, including any whose names I may have overlooked. I would like to add that I have undoubtedly made at least my usual quota of mistakes, even with all the help I have received from so many outstanding scholars. I am of course responsible for all errors of any kind that remain.

Finally, I thank most of all my daughter, Ming Beckwith. Uniquely among all those to whom I attempted to explain how the recursive method works when I first figured it out, she not only immediately understood it but also recognized it as essentially the same as the method used today in the laboratory sciences. Having her support was a great comfort. To Mingming, *xiǎogūniǎng*, I dedicate this book.

# ABBREVIATIONS

*E.I.*$_2$: *The Encyclopaedia of Islam*, new edition. Ed. Hamilton A. R. Gibb et al. Leiden: Brill, 1954–2004, and online.

*Taishō*: 大正新修大藏經 Ed. Takakusu Junjirō, Watanabe Kaigyoku, et al. Tokyo: Taishō Issaikyō Kankōkai, 1924–1932, and online.

## Transcription of non-European scripts

### *Arabic*

The standard transcription used by most Arabists today is followed. The consonants are transcribed: ʾ (*ʾalif* and *ḥamza*), *b*, *t*, *th*, *j*, *ḥ*, *kh*, *d*, *dh*, *r*, *z*, *s*, *sh*, *ṣ*, *ḍ*, *ṭ*, *ẓ*, ʿ (*ʿayn*), *gh*, *q*, *k*, *l*, *m*, *n*, *h*, *w*, *y*. The vowels are transcribed: *a*, *ā*; *i*, *ī*; *u*, *ū*.

### *Chinese*

Chinese names, book titles, and similar items are transcribed in the traditional modified Wade-Giles system used by many Sinologists. The Pinyin system is used only for transcription of Chinese syllables or words cited as such.

### *Tibetan*

There is no standard transcription system for any variety of Tibetan. Classical Tibetan is transcribed conventionally here according to a version (Beckwith 1979) of the systems used by scholars who work on canonical Buddhist texts written in Classical Tibetan. The consonants are transcribed: *k*, *kh*, *g*, *ṅ*, *c*, *ch*, *j*, *ñ*, *t*, *th*, *d*, *n*, *p*, *ph*, *b*, *m*, *ts*, *tsh*, *dz*, *w*, *ź*, *z*, ʼ, *y*, *r*, *l*, *ś*, *s*, *h*, – (the last consonant, a glottal stop [ʔ], is normally not transcribed). The vowels are transcribed *a*, *i*, *u*, *e*, *o*. Suffix morphemes of words are written as part of the word, without hyphens, as is a preceding syllable of a compound word when it ends in a vowel; in all other cases, when a syllable in a compound ends in a consonant it is separated from the following syllable by a hyphen.

# Warriors of the Cloisters

## — Chapter One —

# Introduction

The recursive argument method was "the basic vehicle for the analysis of problems in natural philosophy and theology"[1] from the medieval intellectual revolution to the scientific revolution. It was the actual medieval "scientific method," and it is apparently the source of what may be called the "ideal" modern literary scientific method. The origin of the recursive argument method has long been a mystery. Those who have tried to solve it have sought to explain it as an outgrowth of one or more earlier European traditions, but their proposals do not answer the most important questions, so the problem has remained unsolved. The same is true for the history of the college.[2]

This book shows how the recursive argument method, the actual medieval scientific method, was transmitted along with the college to medieval Western Europe from Classical Arabic civilization, and how the Central Asian Muslims had earlier adopted both from Buddhist Central Asian civilization. The recursive argument method is analyzed in detail, and examples are given showing its formation and development at each stage and in each of the relevant languages.

The present chapter attempts to place this topic and related issues, especially the college, in the context of the full scientific culture that developed in medieval Western Europe in connection with the transmission of these two cultural elements.

The difficulty of understanding a complex problem as a whole is well summed up in the folk expression, "You can't see the forest for the trees." It means that one needs to actually leave the forest and stand some distance away from it in order to understand it as a discrete entity—a forest—composed most saliently of trees, and to see that it is different from other discrete entities such as a city, or a mountain range, which similarly are understood as entities from an external vantage point. A complex entity

[1] Grant (2007: 188).

[2] This introductory chapter, like the concluding chapter, is focused on concepts and is intended as an essay, so its annotations are mostly explanatory. For detailed discussion of the history, transmission, examples, and so on, with citations, see the relevant chapters.

with many constituent elements, whether diverse or homogeneous, cannot be comprehended *as a whole* from the inside; it is necessary to step outside it. Similarly, understanding the scientific culture that developed in medieval Western Europe requires a perspective from which it can be seen as a discrete entity. In this particular case, the culture in question was similar to other medieval cultures that had some science of one kind or another, but it was different from them in one essential way: the others did not develop a *full scientific culture*. It is only by comparison with them that the distinctiveness of the Western European development is apparent. This chapter examines the appearance of a full scientific culture in thirteenth-century Western Europe from the external, holistic viewpoint gained by study of the contrasting "control" cases, which are examined one by one in chapter 7.

## The Constituent Elements of a Full Scientific Culture

Western Europeans came into intensive, long-lasting contact with the Islamic world in a very direct and personal way from the beginning of the Crusades in 1096 onward. Leaving aside the purely military aspects of this contact, countless thousands of civilians traveled to the Islamic world as merchants or pilgrims, or both. During this period Western Europeans copied practically every significant nonreligious cultural element of Islamic civilization that they encountered, including Classical Arabic science (the Classical Arabic version of largely Aristotelian "natural philosophy"), Arab-Indian mathematics, the poetry and music of Islamic Spain, and much else. All this is well known.[3]

However, it has long been asserted that the recursive argument method developed independently in medieval Western Europe, although it is unattested in any text composed in Latin before the early thirteenth century. It is traditionally known as the "scholastic method," or *quaestiones disputatae* 'disputed questions' method, and was used in major

[3] It has been continuously known since the Middle Ages, and it has been demonstrated very carefully by scholars on the basis of overwhelming evidence. It is true that some scholars write as if they were unaware of any of this. For example, according to the index, Lindberg (2007) cites "Crusades" only three times in the entire work, all of them trivial or irrelevant. They could be omitted with no effect on his argument; but the book is in many respects a good survey. Other scholars actually still contest the very idea that Arabs, Turks, or Islamic civilization in general could ever have influenced European culture in any significant way. None of this says anything very good about the latter sort of writers. Nevertheless, although the importance of the extensive cultural contact between the Islamic world and medieval Western Europe is accepted here, it is not crucial to my argument, so these problems are left for others.

works of "natural philosophy" (medieval science) and theology by Albert the Great, Thomas Aquinas, Roger Bacon, and many others.[4] The recursive argument method is the "scientific method" of the Middle Ages and Renaissance, when science was mostly not done by experiments—so that one can hardly speak of a regular "experimental method," which did not come to the fore until the scientific revolution—but by public oral or literary disputation. As shown below, the earliest examples of the recursive argument method in Latin are actually found in some of the most famous, important, and influential texts translated into Latin in the mid-twelfth century in Spain from Arabic originals written a century earlier.[5]

Similarly, it has been asserted that the highly distinctive college developed independently in medieval Western Europe, despite the fact that no one has ever been able to find any convincing native roots for the college in earlier medieval Europe or classical antiquity. Yet it is unquestioned that the first college in Western Europe was founded in Paris by a man who had just returned from Jerusalem, a landlocked city. That means he had necessarily traveled overland through some area of the Near East—which already was overwhelmingly Islamic—in order to get to Jerusalem. Most pilgrims and warriors traversed at least part of Syria, where an absolutely identical Islamic institution, the *madrasa*, was widespread by that time. It is also well known in Islamic studies that this institution had first apppeared several centuries earlier in Islamic Central Asia. And it has long been known that an identical institution, with identical functions and the same highly distinctive architectural form, existed already in pre-Islamic times in Buddhist Central Asia, having developed there slowly and organically out of earlier local forms.[6]

That these two cultural elements—the recursive argument method and the college—appeared at the same time in Western Europe, amid a great deal of acknowledged borrowing from the Islamic world during the period of the Crusades, would seem to be sufficient cause for scholars to look to borrowing as their source, too. However, with very few exceptions, that has not been the case, and the Islamic world has been resolutely ignored in connection with them.

The path that other major constituent elements of science followed *before* their transmission from the Islamic world to medieval Western Europe has also been muddied for a very long time. Moreover, it is

[4] This is discussed in detail in chapter 2.

[5] For details and citations, see chapter 6 and appendix A.

[6] For details and citations, see chapter 3.

remarkable that the Central Asian origin of most of the leading natural philosophers of the great age of Classical Arabic civilization has been generally overlooked.

Similar unclarity is widespread in Islamic studies. From the early Middle Ages onward it has been continuously known that the major works of Indian science were translated into Classical Arabic in the late eighth century, well before the flood of translations from Classical Greek began.[7] Indian (and Buddhist Central Asian) thought had a formative influence on early Islamic civilization. Why then do Islamicists continue to argue that Classical Arabic intellectual civilization developed almost exclusively under Graeco-Roman influence?

Medievalists have known for at least a century that the recursive argument method first appears in Latin texts at the very beginning of the thirteenth century, having no known antecedent in earlier Latin or Greek literature. So why do so many medievalists, ignoring the data and the scholarship on these topics, continue to assert that these elements of medieval science developed purely internally in Europe?

Similarly, it has been known for many decades, from literary and archaeological sources, that the Islamic college, or *madrasa*, is simply a "converted" Buddhist college, or *vihāra*. Why then do Islamicists continue to argue that the *madrasa* developed purely internally, under Graeco-Roman or even Persian influence, in the Islamic world?

As remarked above, it is well known that medieval Western Europe came under massive cultural influence from Classical Arabic civilization during the period of the Crusades, and that Europeans borrowed all of the other essential elements of a full scientific culture, along with many other things, from the Islamic world, mostly via Spain. So why did Europeans *not* borrow the recursive argument method and the college too? Those who advocate a native European origin for the college and the recursive argument method would have us believe that Western Europeans eagerly borrowed everything else from Classical Arabic civilization, but *not* these two particular elements, which are the key elements of a full scientific culture. Well, why not? Since they existed in the Islamic world, with which Western Europe was then in intensive contact, the

[7] Beckwith (2009). Gutas (1998: 28ff.) completely omits the era of Hārūn al-Rashīd (r. 786–809) and the translation of Indian scientific works into Arabic at that time. Druart (2003: 98), citing Gutas (1998), remarks that "Syriac and Persian sources are not to be ignored," but "the great translation movement at the time of the early Abbasids certainly concentrated on Greek texts." However, neither notes that, with a tiny number of exceptions, the translation of Graeco-Roman scientific works got under way only in the ninth century. Cf. chapter 5, note 27.

usual argument forces us to imagine medieval Western Europeans saying to themselves triumphantly, "We see that the Muslims have the recursive argument method and the college, but we shall *not* copy them! We shall brilliantly invent precisely the same complex cultural constructs all by ourselves! It will be a pure coincidence!" Since all of the other significant elements of a full scientific culture were borrowed from Classical Arabic civilization, it goes far beyond reasonable doubt to expect any sensible person to believe that medieval Western Europeans just happened to suddenly and *independently* invent complex cultural constructs that were purely *coincidentally* precisely the same as the earlier recursive argument method and college long possessed by the neighboring culture. Surely historians, above all, should hesitate to believe in so many miraculous coincidences and other marvelous exceptions to the normal course of events in the world.

The late George Makdisi, a medievalist knowledgeable in both Arabic and Latin sources, courageously proposed that both the "scholastic method" and the college were borrowed from the Islamic world. Unfortunately, despite some convincing arguments, he was unable to support either proposal well enough to gain acceptance. Nevertheless, some decades ago the origin of the Islamic *madrasa* 'college' itself became firmly known from archaeology, and a few years ago the present writer found actual textual evidence of the transmission of the recursive argument method from the Islamic world to Western Europe.[8]

As shown in this book, the recursive argument method was neither a native European invention nor an isolated borrowing. It came in from the Islamic world along with the college, translations of Classical Arabic scientific works, and translations of Classical Arabic commentaries on Aristotle and works in the tradition of Aristotelian natural philosophy. The method was immediately integrated into the existing Western European tradition of *Sentences* or *Questions*, literary argument formats that list different positions on problems in theology, law, and other fields.

The sudden influx of cultural borrowing from the Islamic world was fundamental to the development of science in Medieval Latin civilization. Rega Wood writes that "James of Venice's translations [directly from the Greek] had been available since about 1150, but Aristotelian analytics, metaphysics, and natural philosophy" had very little impact in Western Europe until after "the Michael Scot translations [which were

[8] Beckwith (2010b), a short summary of some of the salient points. It is superseded by the present book.

accompanied by Arabic commentaries] became available around 1225." The revolutionary changes in European science occurred not because of the appearance of "a series of isolated works by Greek authors" but because of the translations from the Islamic "tradition of Aristotelian natural philosophy" that framed many of the problems that became "central to scholastic natural philosophy. . . . This is true for topics in metaphysics and epistemology and in psychology and biology."[9] The "sciences of metaphysics and meteorology, physics and chemistry, biology and pyschology were introduced together with Arabic Aristotelianism, and it is difficult to imagine what shape they would have taken without that foundation." Without the contributions of Muslim natural philosohers such as Avicenna and Averroës, "comprehensive scientific views of the cosmos focused on significant physical problems might not have arisen in the Latin West."[10] In short, the insights of great Classical Arabic writers are ultimately what got Europeans so excited about science.

The salient elements of the new scientific culture complex that developed in Western Europe in the late twelfth and thirteenth centuries all existed in the Islamic world—the works of Aristotle and other ancients, Classical Arabic works on Aristotelian natural science, Indian mathematics, the recursive argument method, and the college[11]—but because the latter two elements were not integrated into Islamic culture, a full scientific culture, or scientific culture complex, did not develop there, and eventually science declined and largely disappeared. In fact, Aristotelian science was considered a "foreign," non-Islamic science and was viewed with suspicion even during its heyday there,[12] while the recursive argument method and the college were used in Islam almost exclusively for religious purposes, not scientific ones. By contrast, when Western Europeans borrowed these things, they put them to use right away in the pursuit of science, and European culture changed to accomodate them. This accomodation was not superficial or grudging, as science is today in some cultures where it has been adopted or retained for purely practical reasons. With the help of the Arabic commentaries, Western Europeans understood what was exciting about "Aristotelian" science and enthusiastically accepted all these alien cultural elements. Moreover, despite

[9] Wood (2010: 248).

[10] Wood (2010: 265).

[11] But not the *universitas*, a native European institution that began merging with the college shortly after the latter was introduced, as discussed below.

[12] Cf. Grant (1996: 176–186). On the problem of the apparent decline of science in Classical Arabic civilization, see chapter 7.

popular belief, the Church did not suppress science; on the contrary, its success was due mainly to the support of the Church.[13] The focus of Western European intellectual culture thus shifted emphatically to Graeco-Arabic natural science, which was made the official curriculum of the new college-universities by the middle of the thirteenth century.[14]

Although this medieval intellectual revolution affected all aspects of Western European culture in the twelfth and thirteenth centuries, it was above all about science. As Edward Grant has shown, medieval science laid "the foundations of modern science." The scientific revolution "could not have occurred in Western Europe in the seventeenth century if the level of science and natural philosophy had remained what it was in the first half of the twelfth century, that is, just prior to the translation of Greco-Arabic science that was under way in the latter half of that century. Without the translations, which transformed European intellectual life, and the momentous events that followed from them, the Scientific Revolution in the seventeenth century would have been impossible."[15] Moreover, it is absolutely clear that medieval science—and even more so, a full scientific culture—was not the natural outcome of a gradual, centuries-long evolution from the late Roman Empire and the early Middle Ages.[16] "The Greco-Arabic science that entered Western Europe in the twelfth century was not merely the enrichment of a somewhat less developed Latin science. *It signified a dramatic break with the past and a new beginning.* Logic, science, and natural philosophy were henceforth institutionalized in the newly developed universities."[17] The new learning quickly became "a torrent of new ideas" that "radically altered" the intellectual life of Western Europe.[18] It brought about an intellectual revolution there that was centered on science.

[13] See, among others, Grant (1996) and Lindberg (2007).

[14] See chapters 3 and 6.

[15] Grant (1996: 170–171); cf. Wood (2010).

[16] For Western European scientific endeavors in the early Middle Ages, see Eastwood (2002) and McCluskey (1998); cf. Lindberg (2007: 194ff.), who, however, does not distinguish early medieval scientific efforts, such as they were, from those of the thirteenth century, which depended more or less completely on the transmission of Classical Arabic science to Western Europe. The fact that the unique scientific achievements of Gerbert of Aurillac (ca. 945–1003) were due to his precocious acquisition of Classical Arabic works on a journey to Catalonia in 967 (Lindberg 2007: 201) emphasizes how complete was the difference between the early Middle Ages and the intellectual revolution of the late twelfth and thirteenth centuries.

[17] Grant (1996: 205), emphasis added.

[18] Lindberg (2007: 215). Haskins (1927) and many others after him refer to the twelfth-century intellectual ferment as a "renaissance."

Most of the essential elements in the development of medieval science and a full scientific culture have been treated quite well by previous scholars, most recently and clearly by Grant. However, the structure and lineage of the recursive argument method has hitherto been unclear, and it has not been shown how it and the college were borrowed by Latin Europe from Classical Arabic civilization at about the same time, nor that they originally came from beyond the Islamic world, nor why these particular elements should have been so decisive in establishing the scientific culture of Western Europe.

It is shown in this book that the original, native cultural context or home of the recursive argument method and the college was not in the Near East or the Islamic world at all, but in ancient Buddhist Central Asia. These two constituent elements were the determining factors that distinguish earlier science from a full scientific culture. Although the very same factors had become part of Islamic civilization when Central Asia converted to Islam, they never became crucially important to Classical Arabic *science*, which eventually declined drastically. But when the same elements were transmitted to Western Europe, they became of central importance in Medieval Latin culture and were responsible for the development of the world's first full scientific culture. It is this culture that led eventually to the scientific revolution of the Enlightenment, the direct forerunner of modern science.

The current scholarly consensus is that the practical differences between actual medieval science and actual Enlightenment science (or early modern science), and between the latter and contemporary science, are substantial, regardless of terminology. However, the word *science* in its contemporary usage is fully "modern,"[19] and it is also undeniable that the Medieval Latin term *philosophia* had a very different field of reference[20] from that of the Modern English term *philosophy* and its equivalents in other modern European languages. The goal and methods of Aristotelian science are what al-Ghazālī (Algazel) argues against in his famous book translated into Latin as *Destructio philosophorum*, which

[19] My usage of the terms *science, scientific method*, and others in this book mostly agrees with the usage of Grant (2007: 319ff.), perhaps the world's leading historian of medieval Western European science. Though I have tried to avoid apparent equivalences such as *philosophia naturalis*, which is usually inaccurately converted into English as 'natural philosophy', I have found the latter term useful to refer—as it frequently does in practice—to Aristotelian science, with its heavy philosophical (especially metaphysical) bent.

[20] Or rather fields of reference; the term *philosophia* itself did not by any means always encompass exactly the same things.

does not really mean 'The Destruction of the Philosophers' in Modern English but rather more like 'The Destruction of the Scientists', taking 'science' here in its Aristotelian sense, in which metaphysics is considered to be the most important of all the sciences comprised by *philosophia naturalis* 'natural philosophy'.

Grant's discussion of the meanings of *science* from what may be called the medieval intellectual revolution through the early modern scientific revolution up to today concludes by asking whether medieval science was concerned with problems we now would call "scientific." In reply he gives examples of typical problems or "questions" treated by medieval scholars in works written according to the recursive argument method. Categorizing them according to modern scientific fields, they include:[21]

***Physics***

- "whether, in local motion, velocity is measured according to distance traversed"
- "whether a ratio (*proportio*) of velocities in motions varies as a ratio of ratios of the motive powers to the resistances"
- "whether every visual ray is refracted in meeting a denser or rarer medium"

***Geology***

- "whether the waters of springs and rivers are generated in the concavities of the earth"
- "whether the tranquility of the air is a sign of the earth's motion [i.e., an earthquake] to come"

***Chemistry***

- "whether elements remain [or persist] formally in a compound [or mixed] body"
- "whether a compound (*mixtio*) is natural"

***Meteorology***

- "whether the middle region of air is the place where rain is generated"
- "whether thunder is caused by fire extinguished in a cloud"

[21] Grant (2007: 236–237), q.v. for the source of each *quaestio* 'question'.

The answers to these questions might be known today, but the questions are certainly what we would call "scientific." Although the word *science* did not acquire its modern sense until the nineteenth century, it is clear that what medieval scientists were doing conceptually, however different from what modern scientists do today, was essentially *science* in the contemporary common-parlance sense. Because the categories included in the medieval term *philosophia naturalis* do not correspond exactly to those included in the modern sense of *science*, in order to talk about medieval "science" it is still necessary to use the term *science* in some form or other.[22] As for the "scientific method" used by medieval scientists, it was the recursive argument method, the main topic of this book.

[22] The commonsense approach of Grant (2007: 234ff.) is followed in this book.

## — Chapter Two —
# The Recursive Argument Method of Medieval Science

The distinctive argument method used in scientific literature from the High Middle Ages to the Enlightenment was the "scientific method" until the scientific revolution. It is traditionally referred to in earlier scholarly literature as the 'scholastic method' or *quaestiones disputatae* 'disputed questions' method.[1] Unfortunately, because of increasing scholarly confusion about the origins and meaning of the traditional term *scholastic method*, and even of the term *quaestiones disputatae*, it has been necessary to adopt a purely descriptive term, namely *recursive argument method*, or more briefly, the *recursive method* or *recursive argument*. All refer to the same thing: the highly distinctive argument structure used in the Medieval Latin summas and other works by Robert of Courzon, Alexander of Hales, Albert the Great, Thomas Aquinas, Roger Bacon, and very many others, including before them the Central Asian Muslim philosopher Avicenna, and before him, in turn, Vasubandhu and other Central Asian Buddhist scholars.

The distinctiveness of the recursive argument method, under one name or another, has been recognized by medievalists for well over a century. Although much medieval and Renaissance natural philosophy was written using the recursive argument method, 'natural philosophy' refers to a field of study. The recursive argument is content-free. It was used for works in many fields, including theology and natural philosophy, but also

[1] Moore and Dulong (1943: xv), Wippel (1985: 163), Grant (1996, 2007), and others still understand the traditional sense of *scholastic method* and *disputed questions*. Unfortunately, the substantial literature on the European method—including the above works, as well as Lawn (1993), etc.—is overwhelmingly devoted to attempting to find a "native" source for it. The "classic" recursive argument method is therefore typically mixed up with an earlier nonrecursive format going back to antiquity usually called simply *sententiae* 'sentences; credible views' or *quaestiones* 'questions', and some writers have enlarged the semantic scope of the term *scholastic method* to include practically anything written in Medieval Latin. The new term, *recursive argument method*, which is explained in this chapter, is illustrated with further examples in chapters 4, 5, 6, and 7, and appendix B. Its use in oral disputation is described by Warichez (1932: xlv) on the basis of actual texts. The clarification of its nature, structure, and history in this book obviates the need to further discuss confused notions such as that there were many "scholastic methods," that the "scholastic method" is to be equated with "scholasticism" in general (whether theology, philosophy, science, or argumentation), and so on.

many other fields. From the thirteenth century to the late Renaissance it was the most important literary form for scientific works and was of fundamental importance for medieval science and the history of science in general.[2]

First of all, it must be clearly distinguished from other argument forms.

## Treatise Argument Structure

Most major ancient, medieval, and modern scientific works, including the present book, are treatises. Typical treatises have little or no overt or explicit argument structure, but simply begin with the topic (usually mentioned in the title) and proceed, in the author's voice, from the beginning to the end of the text.[3] Texts written as traditional treatises, such as the works of Aristotle and most modern scholarly works, have essentially no overt, explicit argument structure.[4] The basic structure of the treatise may be summarized as: T : AAAAAAAAAA . . . (T = topic; A = author's view), or as follows:

*Title of Book*

> The author's view begins and proceeds to discuss the topic, with occasional comments on other views. The length of the work is irrelevant, but if it is very long, it is usually broken up into chapters and sections, purely for the sake of convenience. This is the structure of Aristotle's works, most of the Muslim Aristotelians' works, and earlier medieval works in general. The point is to present information to the reader. Any 'objections' or references to views with which the author disagrees are presented incidentally within the author's narrative. The treatise has no particular overt, external, explicit formal structure other than that determined by the topics of interest to the author, who can present them in any order he pleases. Like this paragraph, it continues on until the end, and then simply stops.

[2] Grant (2007: 188, 236ff.).

[3] Physical division of a text into scrolls, bundles of bamboo slips, etc., also typically has nothing to do with the overt, explicit argument structure of the text per se, though such division often has something to do with the *content* of the text.

[4] Medieval authors often use the treatise format instead of the scholastic method, but the latter was by far the most widely used for *philosophia naturalis* (Grant 2004: 175). Cf. Grant (2004: 172ff.; 2007: 182–183) on the treatise, commentary, and recursive argument method (his '[disputed] question' format).

Many medieval scholars who wrote works using the recursive argument method also wrote treatises. For example, in the *Treatise on Proportions* or *On the Proportions of the Speeds of Motion*, Thomas Bradwardine (ca. 1290–1349) follows this structure. The short précis of the entire work given in his introduction may be analyzed as follows:[5]

***Title:*** On the Proportions of the Speeds of Motion

***Introduction,*** a very brief preface on the premises, approach, and general topic of the book.

***Chapter 1,*** "setting forth the necessary mathematics," *is divided into three parts*, the first on "the definitions, types, and other properties of proportion," the second on "proportionality," and the third on "certain axioms, from which several mathematical conclusions are drawn."

***Chapter 2,*** arguing against opinions on "the proportion between the speeds of motion and following the number of those opinions," *is divided into four parts.*

***Chapter 3,*** presenting "the correct understanding of the proportion between the speeds of motion" *is divided into two parts.* "The first of these develops several theorems concerning the proportion between the speeds of motions, and the second raises and settles objections to them."

***Chapter 4,*** covering further aspects of motion, *is divided into three parts*, the first "establishing the requisite mathematical material," the second refuting "several opinions concerning the proportion between the speeds of motions," and the third on "certain hidden truths concerning the proportions between the elements."

## Dialogue Argument Structure

Texts written in the traditional dialogue or question-and-answer format are also completely different from the recursive argument method. The "Socratic dialogue" format followed most famously in the works of Plato is actually a covert treatise, as are other such continuous two-voice structures. Despite the presence of an overt second voice, the latter is rarely

[5] This analysis summarizes the presentation in Grant (2007: 188–190); cf. Grant (2004: 172–174).

presented as a genuine alternative to the view of the authorial voice, but even if a genuine objection is raised it is promptly refuted. For example, consider the *Dialogue Concerning the Two Chief World Systems* by Galileo Galilei (1564–1642), in which the dialogue's participants, here Salviati (SALV.) and Sagredo (SAGR.), argue about different theories of the solar system:[6]

SALV. Taking up again what you granted me a short time ago, that the new star could not be in more than one place, then whenever the calculations made from the observations of these astronomers do not agree in putting it in the same place, there must be errors in the observations; that is, either in taking the elevation of the pole or the altitude of the star, or both. Now, since in many estimates, made from the combinations of the observations two at a time, there are a few which place the star in the same position, only these few can be free from error; . . .

SAGR. Then one would have to trust these few alone more than all the rest put together. And since you say that there are few of these which agree, and I see two among these twelve which put the distance of the star from the center of the earth at four radii (the fifth and sixth of them), then the star is more likely to have been elemental. . . .

SALV. Not so, for if you will look carefully, it does not say here that the distance is exactly four radii, but about four radii. And you see that those distances differ between themselves by hundreds of miles. Look here: this fifth one, you see, which is 13,389 miles, exceeds this sixth one of 13,100 miles by nearly 300. . . .

SAGR. Then which are these few which agree in placing the star in the same position?

SALV. There are five investigations, to the disgrace of this author, which all place it among the fixed stars, as you may see in this other note where I have recorded many more combinations. . . .

The overt structure of the dialogue argument is clearly: T : A(B)A(B)A(B)A(B)A(B)A . . . , but its covert (actual) structure is T : AAAAAAAAAA . . . or simply: T : A. The treatise and dialogue argument types are thus essentially one type underlyingly.

[6] Galilei (1967: 336–337).

## Recursive Argument Structure

The contrast between the treatise and dialogue argument structures, on the one hand, and the recursive argument, on the other, could hardly be more striking. Yet even more remarkable is the structure of an entire book written according to the recursive argument method, such as the summas of the great thirteenth-century philosopher-theologians. Such a book does not consist of a continuous authorial viewpoint, either overtly as in an Aristotelian treatise or covertly as in a Socratic dialogue. Instead, it is broken up into very many full recursive arguments ('disputed questions') on specific points.

A book written according to the recursive argument method typically has a general, encyclopedic title such as *Summa . . . X* 'Summary of [Everything on the Subject of] X', or *Quaestiones Disputatae . . . Y* 'Disputed Questions on [the Subject of] Y'. After an optional short preface or prologue, the text begins with the first argument, the topic of which (typically its title) is an argument usually presented as a statement or question about the interpretation of a quotation from scripture or of other literature relevant to a point of theological doctrine, or about a problem in one of the sciences. This MAIN ARGUMENT (the QUESTION proper or TOPIC) is followed by the presentation of an argument (SUBARGUMENT$_1$), or normally a string of arguments, the SUBARGUMENTS$_1$, about it. In the SUBARGUMENTS$_1$ section, variant views by other writers, as well as hypothetical arguments that could be made, are presented. This first string of arguments is followed by a second string of arguments, the SUBARGUMENTS$_2$, about the arguments in the first string,[7] presented in the same order. This time, definitive arguments for or against the arguments in the first list (the SUBARGUMENTS$_1$) are presented. The SUBARGUMENTS$_1$ are typically identified by number in the SUBARGUMENTS$_2$ list, but some authors do not number them, referring to each SUBARGUMENT$_1$ only by a brief summary statement of its main point at the beginning of its corresponding SUBARGUMENT$_2$, or by mentioning the author or textual source of the SUBARGUMENT$_1$ to which the writer is responding.[8] The AUTHOR'S VIEW ARGUMENT—a necessary part of the full recursive argument method—may occur in different locations within a full argument,

[7] In modern studies, the subarguments in the first list are often called 'objections', and those in the second list 'rebuttals', or the like, but in fact both lists of subarguments may include views with which the author agrees.

[8] In some of the earliest examples of the method, the subarguments in the first list are referred to in order and by number in the second list. See chapter 4.

depending on the individual writer or the traditional arrangement in his literary culture. The structure of an *individual* SUBARGUMENT (within a full recursive argument) can follow treatise, dialogue, or recursive argument structure. In the latter case, there are recursive arguments within recursive arguments.

After going through this procedure—MAIN ARGUMENT (QUESTION or TOPIC argument), first list of SUBARGUMENTS, AUTHOR'S VIEW ARGUMENT, second list of SUBARGUMENTS—the author then moves on to the next TOPIC, which is treated in exactly the same way (MAIN ARGUMENT, first list of SUBARGUMENTS, AUTHOR'S VIEW ARGUMENT, second list of SUBARGUMENTS). The writer does this topic by topic, in one recursive argument after another, throughout the entire text, which typically consists of many dozens or hundreds of such topics, most of which are presented as full recursive method arguments.

The essential structure of a basic recursive argument is thus:

$$\mathrm{T} : \mathrm{W_1 X_1 Y_1 Z_1} : \mathrm{A} : \mathrm{W_2 X_2 Y_2 Z_2}$$

In this structure, T is the MAIN ARGUMENT or TOPIC,[9] A is the AUTHOR'S VIEW ARGUMENT (shown in its default position in Medieval Latin texts),[10] and W, X, Y, and Z are the SUBARGUMENTS.

### *Basic Recursive Argument Method Structure*

#### *Title of Book*

1. First Recursive Argument

I. MAIN ARGUMENT (TOPIC / TITLE / QUESTION)
II. SUBARGUMENTS$_1$ *Pro* and *Contra* the argument in I

[1] SUBARGUMENT$_1$ *Pro* or *Contra* the argument in I
[2] SUBARGUMENT$_1$ *Pro* or *Contra* the argument in I
[3] SUBARGUMENT$_1$ *Pro* or *Contra* the argument in I
[4] SUBARGUMENT$_1$ *Pro* or *Contra* the argument in I
[5] SUBARGUMENT$_1$ *Pro* or *Contra* the argument in I

[9] The MAIN ARGUMENT (or TOPIC) is usually phrased as a question: "It is asked whether . . . " In some Medieval Latin recursive argument method books, the topics are listed in a summary introduction to each chapter and are referred to only by number at the beginning of each full argument. It is common for more than one question to be raised in the MAIN ARGUMENT. Other variations also occur. In an example from Avicenna's *De anima* analyzed in chapter 6, the author also gives an additional initial listing of the subarguments, evidently as a sort of table of contents.

[10] See above on the location of this element.

III. AUTHOR'S VIEW ARGUMENT[11] PRO or *Contra* the argument in I
IV. SUBARGUMENTS$_2$ about the SUBARGUMENTS$_1$ listed in II

[1] SUBARGUMENT$_2$ *Pro* or *Contra* SUBARGUMENT$_1$ [1] in II
[2] SUBARGUMENT$_2$ *Pro* or *Contra* SUBARGUMENT$_1$ [2] in II
[3] SUBARGUMENT$_2$ *Pro* or *Contra* SUBARGUMENT$_1$ [3] in II
[4] SUBARGUMENT$_2$ *Pro* or *Contra* SUBARGUMENT$_1$ [4] in II
[5] SUBARGUMENT$_2$ *Pro* or *Contra* SUBARGUMENT$_1$ [5] in II

## 2. Second Recursive Argument

I. MAIN ARGUMENT (TOPIC / TITLE / QUESTION)
II. SUBARGUMENTS$_1$ *Pro* and *Contra* the argument in I

[1] SUBARGUMENT$_1$ *Pro* or *Contra* the argument in I
[2] SUBARGUMENT$_1$ *Pro* or *Contra* the argument in I
[3] SUBARGUMENT$_1$ *Pro* or *Contra* the argument in I

III. AUTHOR'S VIEW ARGUMENT *PRO* or *Contra* the argument in I
IV. SUBARGUMENTS$_2$ about the SUBARGUMENTS$_1$ listed in part II

[1] SUBARGUMENT$_2$ *Pro* or *Contra* the SUBARGUMENT$_1$ [1] in II
[2] SUBARGUMENT$_2$ *Pro* or *Contra* the SUBARGUMENT$_1$ [2] in II
[3] SUBARGUMENT$_2$ *Pro* or *Contra* the SUBARGUMENT$_1$ [3] in II

## 3. Third and Following Recursive Arguments

These follow the same structure as the first and second recursive arguments above.

In Vasubandhu, and typically in Medieval Latin works, the subarguments are grouped or subdivided in one or more ways, such as arguments according to scripture and according to reason, or arguments *Pro* and arguments *Contra* the MAIN ARGUMENT. The latter type of subdivision is normally reflected in the recursion, in that any subarguments of the first list with which the author agrees are regularly omitted in the second list of subarguments—that is, the author does not argue against positions with which he agrees. This subvariety of the recursive argument method is structured as follows, marking the AUTHOR'S VIEW A with a subscript P for *Pro* or C for *Contra*:

[11] This is the typical position of the AUTHOR'S VIEW ARGUMENT in Latin texts.

$$T: W_P X_P Y_C Z_C : A_C : W_P X_P$$

*or:*

$$T: W_C X_C Y_P Z_P : A_P : W_C X_C$$

In this widespread variation of the method, the author responds in the SUBARGUMENTS$_2$ section only to the SUBARGUMENTS$_1$ with which he disagrees. (The overt formatting of this variant in Medieval Latin texts nevertheless follows that of the "basic" method, so the AUTHOR'S VIEW ARGUMENT is retained as a major structural category.) It may be called the "parsimonious" variant. Because the AUTHOR'S VIEW ARGUMENT (in general and in this variant) is actually the last of the SUBARGUMENTS$_1$, the parsimonious variant has either of the two following *actual* structures:

First Version: Author's View = SUBARGUMENTS$_1$ *Contra*

I. MAIN ARGUMENT (*Quaestio*)
II. SUBARGUMENTS$_1$ about the MAIN ARGUMENT in I

  a. SUBARGUMENTS$_1$ *Pro*
  b. SUBARGUMENTS$_1$ *Contra*

    α. AUTHOR'S VIEW ARGUMENT *Contra* (*Solutio*)

III. SUBARGUMENTS$_2$ about the SUBARGUMENTS$_1$ in II

  a. SUBARGUMENTS$_2$ *Contra* the SUBARGUMENTS$_1$ *Pro*
  b. AUTHOR'S VIEW ARGUMENT *Contra* (*Determinatio*; optional)

Second Version: Author's View = SUBARGUMENTS$_1$ *Pro*

I. MAIN ARGUMENT (*Quaestio*)
II. SUBARGUMENTS$_1$ about the MAIN ARGUMENT in I

  a. SUBARGUMENTS$_1$ *Contra*
  b. SUBARGUMENTS$_1$ *Pro*

    α. AUTHOR'S VIEW ARGUMENT *Pro* (*Solutio*)

III. SUBARGUMENTS$_2$ about the SUBARGUMENTS$_1$ in II

  a. SUBARGUMENTS$_2$ *Contra* the SUBARGUMENTS$_1$ *Contra*
  b. AUTHOR'S VIEW ARGUMENT *Pro* (*Determinatio*; optional)

## *Parsimonious Recursive Argument Method Structure*

### *Title of Book*

### 1. First Recursive Argument

I. Main Argument (Topic / Title / Question)

II. Subarguments$_1$[12] *Pro* and *Contra* the argument in I

- [1] Subargument$_1$ *Pro* the argument in I
- [2] Subargument$_1$ *Pro* the argument in I
- [3] Subargument$_1$ *Pro* the argument in I
- [4] Subargument$_1$ *Contra* the argument in I
- [5] Subargument$_1$ *Contra* the argument in I

III. Author's View Argument *Contra* the argument in I

IV. Subarguments$_2$ about the Subarguments$_1$ listed in II[13]

- [1] Subargument$_2$ *Contra* Subargument$_1$ [1] *Pro*
- [2] Subargument$_2$ *Contra* Subargument$_1$ [2] *Pro*
- [3] Subargument$_2$ *Contra* Subargument$_1$ [3] *Pro*

### 2. Second Recursive Argument

I. Main Argument (Topic / Title / Question)

II. Subarguments$_1$ *Pro* and *Contra* the argument in I

- [1] Subargument$_1$ *Contra* the argument in I
- [2] Subargument$_1$ *Contra* the argument in I
- [3] Subargument$_1$ *Pro* the argument in I

III. Author's View Argument *Pro* the argument in I

IV. Subarguments$_2$ about the Subarguments$_1$ listed in part II

- [1] Subargument$_2$ *Contra* Subargument$_1$ [1] *Contra*
- [2] Subargument$_2$ *Contra* Subargument$_1$ [2] *Contra*

### 3. Third and Following Recursive Arguments

These follow the same structure as the first and second recursive arguments above.

[12] Part II can contain any number of subarguments, but it may contain only one, in which case part IV will have only one subargument (or even none if the author agrees with it), as in Example 6.3 by Avicenna in chapter 6.

[13] Most Latin authors use the parsimonious recursive argument method and regularly omit the Subarguments$_2$ that would correspond to the view or views accepted by the author.

As an example of the parsimonious version of the recursive argument method, the one typically used by Medieval Latin scholastics, consider EXAMPLE 2.1, from the *Quaestiones super* 'De animalibus' of Albertus Magnus (Albert the Great). It is a scientific work treating problems raised by Aristotle's zoological treatise *De animalibus* 'On animals'. It consists more or less exclusively of recursive arguments from start to finish. Albert gives his own AUTHOR'S VIEW ARGUMENT in the middle and again at the end.

### *Example 2.1*

**ALBERTUS MAGNUS (Albert the Great)**
***Quaestiones super* 'De animalibus'**

BOOK 1, QUESTION 3
Whether an Organic Member That Has Been Cut Off Can Be Restored[14]

[I. MAIN ARGUMENT (TOPIC OR QUESTION)]

It is inquired[15] whether an organic member that has been cut off can be restored.

[II. SUBARGUMENTS$_1$]

[a. SUBARGUMENTS$_1$ (*Pro*)]

[1] It appears to be so[16] [L. *Videtur quod sic*], because in animals organs are as branches in plants; but a branch that has been cut can be regenerated. Therefore, by the same reason [so can] organic members in animals.[17]

[2] Again [L. *item*], organic members are composed of similar parts; but similar parts can be regenerated, as is seen in flesh. Therefore . . . , and so on.

[3] Again [L. *item*], nutriment is converted into the substance of the one nourished. Therefore, it is possible that any member that is lost can be restored by nutriment.

[14] Latin text: Filhaut (1955: 80–81). The English translation is mostly from Grant (1974: 681–682). In a few places—all minor (and noted)—my translation is substituted for Grant's. I have also added subsection labels [in brackets], according to my analysis of the recursive method in this chapter, and have made minor changes in the punctuation. Note that here, as elsewhere, all bracketed additions are my attempts to help clarify the text.

[15] Grant (1974: 681) has: "We inquire . . . "

[16] Grant (1974: 681) has: "It seems that it can . . . "

[17] Grant (1974: 681) has: "Therefore, by the same reason, organic members in animals [can be regenerated]."

[b. SUBARGUMENTS$_1$ (*Contra*)]

[4] The opposite [L. *oppositum*][18] is obvious to the senses; a hand that is cut is not regenerated, nor an eye that has been torn out.

[III. AUTHOR'S VIEW ARGUMENT (*Contra*)]

It must be said that [L. *dicendum, quod*], organic or functional parts in animals cannot be restored, because in proportion as a thing is more noble so nature is more concerned about its production. . . . [One paragraph]

[IV. SUBARGUMENTS$_2$] [Response] to the [principal] reasons.[19]

[a. SUBARGUMENTS$_2$ *Contra* the SUBARGUMENTS$_1$ *Pro*)]

[1] To the first, it should be said that branches that have been cut can be regenerated . . .

[2] To the second, it should be said that certain parts, as nerves, bones, and similar [or homogeneous parts] (L. *consimilia*) cannot be restored any more than organic parts. . . .

[3] To the third, it should be said that although nutriment can be converted into the substance of the one who is nourished, nevertheless, when a hand has been cut off, the power which might convert the nutriment into a likeness of the hand is lacking.

[V. AUTHOR'S VIEW ARGUMENT (CONCLUSION)]

And so, as has been declared [in SUBARGUMENT$_1$ [4] (*Contra*)] it is plain that such members cannot be restored by nutriment, etc.[20]

Modern readers might think that the viewpoints expressed in the arguments are sometimes hardly "scientific," but that misses the point of the argument structure. The idea was to present all significant views on a disputed subject and to examine them rigorously using the science of the day, which was dominated above all by Aristotelian logic.

[18] Grant (1974: 682) has: "The opposite [of this] is obvious . . . " Note that Albertus regularly uses *oppositum* in this work rather than the usual *contra*.

[19] Latin *rationes* 'explanations, reasons'. The formula Albertus follows in this work is: *Ad primam/secundam/tertiam/etc. dicendum, quod*. That is, although the SUBARGUMENTS$_1$ are not numbered, the replies to them (SUBARGUMENTS$_2$) are numbered. In addition, in the second list he does not mention the objections from the first list with which he agrees.

[20] The "etc." is in the original Latin.

There is considerable variation in the practical use of the method by authors even in the same tradition. Although the title of the chapter or section, when there is one in the original, is usually the same as the main argument or topic, some works are broken into chapters in which the argument topics are listed in a summary introduction to the chapter and then are only referred to by subsection number at the beginning of each recursive method argument.

The AUTHOR'S VIEW ARGUMENT can occur in any one of three positions—before the first list of SUBARGUMENTS, between it and the second list of SUBARGUMENTS, or after the second list of SUBARGUMENTS—depending on the tradition or on the individual author. It often is represented, at least partially, in more than one of the three positions, and even in all of them, regardless of its expected formal position in works of the author's tradition.

In more complicated recursive arguments, an individual subargument (within an individual recursive argument) can itself be structured as a full recursive argument, in which some of the subarguments are themselves structured as full recursive arguments. In all texts examined there are some arguments that do *not* follow the formal recursive method.[21] Also, some argumentative "discussions" (see the example by Avicenna in chapter 6) follow the recursive argument method, but others do not. There are many variations in the argument format even within one author's works, or one particular work, and many variations within the arguments of each tradition.[22]

The most characteristic features of the recursive argument method are its *recursive structure* and *internal lists of arguments* representing views, both real or hypothetical, on a problem. In its rigorous and exhaustive recursive analysis the method is sharply differentiated from all of the earlier "scholastic" works that have been claimed to be its precursors in Western Europe, such as Gratian's *Decreta*, Abelard's *Sic et non*, and other examples going back to antiquity, as discussed below. These works do indeed contain lists of arguments, positions, variant views from Scripture, or whatever, but none consist of *recursive* arguments. Formal recursion is the distinctive, crucial aspect of the method: the arguments in the second list (SUBARGUMENTS$_2$) dispute the arguments in the first list (SUBARGUMENTS$_1$), which dispute the argument in

[21] In addition, even some of the great Latin summas' arguments occasionally do not contain all of the expected sections, while other arguments can be simple treatises.

[22] Each of these deserves careful study.

the MAIN ARGUMENT (the TOPIC or QUESTION), as shown in EXAMPLE 2.1, above.[23]

As mentioned, it has long been established that the recursive argument method does not exist in the works of Aristotle or any other Ancient Greek or Latin writer. It has also not yet been found in the works of Classical Arabic scholars before Avicenna (Ibn Sīnā). The method first appears in Medieval Latin in the twelfth century, but it earlier appeared in Classical Arabic civilization. Moreover, it is now recognized that science in general had already appeared (or in some views, reappeared) in another culture before it appeared in Europe—not just any culture anywhere, but precisely the one that was the main donor of scientific works to Latin Western Europe: Classical Arabic civilization. Some of the works by the most outstanding scientists of that culture were translated into Latin in Western Europe, mostly in Spain.

The translation of Avicenna's *De anima* and *Metaphysics* in the mid-twelfth century accomplished the literary transmission of the recursive argument method to Latin Europe, though it is quite possible, even probable, that the method was also transmitted orally, the way Avicenna himself learned the method in his youth.[24] The Latin scholars of the thirteenth and subsequent centuries, following his lead in commenting and expanding on Aristotle, used the recursive argument method for all Aristotelian topics, which meant all sciences then known.[25] Eventually, every field of knowledge treated in this way automatically became by extension a more scientific one.[26]

It is possible that one aspect of the recursive argument method that appealed to scientific-minded authors, besides formal recursion, is the use of numbering: the second list of subarguments is typically numbered explicitly. Sometimes there are very many subarguments in a list, and some of the subarguments themselves also contain full recursive arguments, which may also be numbered. However, there is considerable variation in how this is actually done. In some authors, none of the subarguments are numbered explicitly, or only the first one or two are numbered. In authors who do regularly number the subarguments, normally

[23] Examples from the Central Asian Buddhist tradition and the Classical Arabic tradition are given in chapters 4 and 5, respectively. Other examples from the Medieval Latin tradition are given in chapter 6.

[24] See chapter 5.

[25] Cf. Grant (1996: 40 and infra).

[26] It is akin in this way to mathematics, which can be put to use for many purposes, both scientific and nonscientific.

only one of the two lists is actually numbered—for example, usually either the first list is numbered or the second list is numbered, but not both—though some, such as Fakhr al-Dīn al-Rāzī, scrupulously number both lists. As discussed above, in many texts the SUBARGUMENTS$_1$ with which the author agrees are omitted in the SUBARGUMENTS$_2$. This occurs already in Vasubandhu, and is found in the summa of Robert of Curzon,[27] thought to be the earliest work authored in Latin in which the recursive argument method is certainly attested. It is also found in the *Sentences* of Peter of Poitiers,[28] in the great Medieval Latin summas, and in many other works.

The general idea of the method is to break each topic down into analyzable parts and exhaustively debate each one from every possible direction. Its practical importance is in the fact that, in using the method, many views, very often including hypothetical ones, are presented and examined on each point, the irrelevant or wrong views are disproved, and the author's view is established firmly.[29]

While this approach would seem to suggest a certain scepticism about arguments in general, and it did eventually have this effect, it does not seem to have happened in the thirteenth century, due to the overwhelming influence of the then "new" Aristotelian system. The two meshed very well, because the point of the recursive argument method and its goal both correspond to what is said by Aristotle to be the point and goal of science: the Truth.

While comparison of specific ideas of Aristotle or other early Greek thinkers with those typical of Sarvāstivāda, the dominant school of Central Asian Buddhism, has hardly begun, it must be admitted at the outset that these systems are radically different. Nevertheless, it is highly improbable that the Greeks and the early Buddhists had *no* influence at all on each other,[30] and even more unlikely that the later Central Asian Buddhists had *no* influence on early Islam,[31] as current scholarship would have us believe.

[27] See chapter 6.

[28] It is not yet possible to date the work or the author. See appendix B.

[29] This is the reason Aristotle gives for the presentation of arguments against one's own view, but again, it must be emphasized that he is totally unconcerned with the rhetorical (external) structure of the argument per se, but only with the logical structure, with practical methods for debating opponents who are interested in winning rather than in finding the truth, and so forth. See, e.g., *Metaphysics* iii, 1; *Topics* viii, 9. These are precisely the same interests of the later Buddhist writings on logic and debate (for example, the works of Dignāga, Dharmakīrtī, and so on), some of which use the recursive argument method type prevalent in India, q.v. chapter 7.

[30] See chapters 3 and 4.

[31] Practical problems include the fact that most major Sarvāstivāda works are preserved only in Chinese translations, which use a specialized, allusive scholastic metalanguage difficult to

## Formal Recursion and the Recursive Argument Method

The recursive argument method represents thinking about one's own thinking, including arguing about one's own arguments, a variety of what is known as linguistic recursion. This method of analysis and disputation developed first of all among Central Asian Buddhist scholars interested in disproving the beliefs of other Buddhist sects,[32] as discussed in chapter 4. The recursive method is at heart a way to examine a problem systematically, logically, and in great detail. The radical, new, rigorous approach to formal analysis represented by the method must have been exciting for the young scholars of Western Europe who first encountered it. No later than the mid-thirteenth century it began to be used in Medieval Latin works devoted specifically to the natural sciences. Albert the Great thus used it in his summa as well as in important scientific works, as did Roger Bacon (ca. 1219–1294)[33] and many others. It quickly became "the most commonly used format for the presentation of natural philosophy at universities" in the Middle Ages.[34]

There are different types of formal recursion, including artistic, literary, linguistic, mathematical, and computational recursion, but they all share the fundamental feature of embedding. One of the best-known types is the artistic example, as in the *matryoshka* doll (or Russian "nesting dolls"), where one doll contains another one, which contains another one, which contains another one. The recursive argument's structure is structurally a subtype of linguistic recursion, but it has a few special twists of its own. Linguistic recursion occurs when clauses are embedded within clauses that are embedded within clauses, most clearly perhaps in "which (or that) clause" sentences, such as: *The bear ate the dog that ate the cat that ate the mouse that ate the cheese.* The recursive argument is,

understand even for those who know Classical Chinese. Moreover, not one of them has ever been published in a modern critical edition, and although a few of the major texts of the Sarvāstivāda Buddhist tradition have been translated into a Western language, those translations that do exist are written in a metalanguage that is nearly incomprehensible to all but the tiny number of specialists who produce the translations. By contrast, the surviving texts of Aristotle are well known because they have been published in modern critical editions and have been translated into many languages for the enlightenment of those who cannot read Greek well enough to understand Greek philosophical texts. Finally, most scholars who have attempted to compare these two systems (or the Islamic and Indian systems) have not been familiar enough with both sides of the traditions they have compared.

[32] Ibn Sīnā, the earliest known author in Classical Arabic civilization in whose works it occurs, uses it partly for quite the same purpose, though the religion is Islam.

[33] His *Questiones supra libros prime philosophie Aristotelis (Metaphysica i, ii, v–x)*, on Aristotle's *Metaphysics*, is entirely written according to the recursive method (Bacon 1930), as is his *Questiones supra libros octo Physicorum Aristotelis*, on Aristotle's *Physics* (Bacon 1935).

[34] Grant (2007: 190).

minimally, an argument that is disputed by an argument that is disputed by an argument, or more simply (but in reverse order), an argument about an argument about an argument. However, the embedded recursions—that is, each argument except the first (the MAIN ARGUMENT, TOPIC, or QUESTION)[35]—can actually consist of any number of arguments (which are referred to here as SUBARGUMENTS), in which case it consists of arguments about arguments about an argument. This feature involves embedding iterative sets. It is the special twist of the traditional recursive argument method and one of its most characteristic features.

A full recursive argument thus begins with a single MAIN ARGUMENT (the TOPIC, QUESTION, or "base" argument), followed by an argument about it, which is the first recursion. This is an argument *section* that can contain from one to many arguments (SUBARGUMENTS$_1$). It is followed by the second argument section, which can maximally contain the same number of arguments as the first SUBARGUMENT section, or minimally as few as zero arguments (SUBARGUMENTS$_2$). However, the first recursion (i.e., the SUBARGUMENTS$_1$ section) normally consists of a string of arguments, sometimes numbered, which must be disputed *in order* in the second recursion (SUBARGUMENTS$_2$) section, where they are often numbered. A further refinement is that arguments given in the SUBARGUMENTS$_1$ section with which the author agrees (the author's view being given in an AUTHOR'S VIEW ARGUMENT somewhere in the full argument) are typically not mentioned in the SUBARGUMENTS$_2$ section, as in the parsimonious version of the method outlined above.

### *The Recursiveness of the Recursive Argument Method*

I. MAIN ARGUMENT (*Quaestio*).

II. The MAIN ARGUMENT is argued about by the SUBARGUMENT(S)$_1$ section, which includes the AUTHOR'S VIEW ARGUMENT (*Solutio*).

III. The SUBARGUMENT(S)$_1$ is or are argued about (in sequential order) by the SUBARGUMENT(S)$_2$ section, which includes an optional AUTHOR'S VIEW ARGUMENT (*Determinatio*).

In other words, because the first part (I) is actually an ARGUMENT, which is argued about by the ARGUMENT in part II, which is argued about by the ARGUMENT in part III, the method is structurally a form of linguistic recursion. However, parts II and III can contain any number of arguments, and if there are a plurality of arguments in the SUBARGUMENT(S)$_1$

[35] However, even the MAIN ARGUMENT occasionally contains more than one argument.

section, they must be replied to sequentially in the SUBARGUMENT(S)$_2$ section. This is one of the most characteristic features of the method.

A recursive argument thus consists of arguments about arguments about an argument, and at the chapter or book level, arguments about arguments about arguments about an argument—the chapter or book topic. As with other types of recursion, such as mathematical or computational recursion, the recursive argument has its own particular features. A computational-type recursive argument would consist of a list of subarguments, one of which subarguments is itself a list of subarguments, one of which subarguments is itself a list of subarguments, and so on, down to the bottom-level "base case," which does not include a further list of subarguments. Typical literary recursion is "a play within a play." And so on. By contrast, the recursive argument method typically includes a *string* of subarguments within an argument, and requires that the first string of subarguments be repeated and processed sequentially in the second string (the subarguments about the subarguments about the argument), in order to complete the recursion. Since the recursions in a recursive argument occur as in linguistic recursion, which might seem to be in reverse order—they constitute an argument (or arguments, the SUBARGUMENT(S)$_2$) about an argument (the SUBARGUMENT(S)$_1$) about an argument (the MAIN ARGUMENT)—in a recursive argument the base case structurally is the MAIN ARGUMENT (I), which is always stated first. The argument sections are not really in reverse order, but in the normal order of linguistic recursion in languages like English, Sanskrit, Arabic, and Latin.[36]

The recursive argument not only has its own highly distinctive form that differs from other traditional forms of recursion but it also differs sharply from all earlier forms of literary argument. Moreover, the lists of sentences or questions in early scholastic works (often bearing *Sentences* or *Questions* in the title) written from antiquity onward, which supposedly constitute the predecessors of the recursive argument, are mostly not arguments at all,[37] let alone recursive ones.

The full recursive method argument requires at least one statement of the author's position (III in the above example), but it can actually occur in any major position after I—that is, in position II, III, or IV, or in two or more of these locations. There are examples of several such variations in the Central Asian Buddhist, Central Asian Arabic, and Medieval

[36] In Chinese, Japanese, Mongolian, Tibetan, Turkish, and many other languages, the normal order of linguistic recursion is the opposite.

[37] See the discussion below and in chapter 6.

Latin forms of the argument. The Indian recursive argument method is a primitive form in which there is no AUTHOR'S VIEW ARGUMENT section, and the $\text{SUBARGUMENTS}_1$ always represent the opponent's position, while the $\text{SUBARGUMENTS}_2$ always consist of the author's arguments against all of the arguments in the first list.

In some Latin examples, a fifth part (V), the *Determinatio* 'Determination'—often a repetition of the AUTHOR'S VIEW ARGUMENT—is given. Also, in some authors, regardless of tradition, part IV (the $\text{SUBARGUMENTS}_2$ section) discusses in order *all* the $\text{SUBARGUMENTS}_1$ and accepts or refutes each one.

*The Indispensible Parts of a Full Recursive Method Argument*

- A statement of the MAIN ARGUMENT, or Topic [T]
- A list of arguments *pro* or *contra* the main argument; = $\text{SUBARGUMENTS}_1$ [B]
- A list of arguments *pro* or *contra* the arguments in [B], with arguments about each; = $\text{SUBARGUMENTS}_2$ [C]
- A statement of the AUTHOR'S VIEW ARGUMENT [A] after at least one of: T, B, C

It must be stressed that the *internal, implicit, logical* structure of the *content* of a recursive argument—that is, the *logical* structure of the point being made—has nothing whatever to do with its purely *external, explicit, formal* or "rhetorical" structure. The terms *recursive argument method, recursive argument*, and *recursive method*, as used in this book (instead of the traditional term *scholastic method*), all refer specifically and only to an argument with the overt, external organizational form of presentation described in detail, with examples, in this book. In his analyses of the structure of the recursive method argument, Grant refers to it as a "questions" type of argument, his shorthand for a 'disputed questions' (*quaestiones disputatae*) argument.

It is established that this highly distinctive overt or external argument structure is unknown in ancient or early medieval Western European authors' works and has no earlier Latin or Greek precedents.[38] However, there are many claims to the contrary. Makdisi follows European medievalists in ascribing the origins of the recursive argument method to Peter Abelard and to still earlier legal scholars, but unlike them, he claims these

[38] Kantorowicz (1938: 58–59)

earlier Latin scholars got the method from Arabic legal disputation via the transmission of the college to Europe. Makdisi thus argues that the Latin recursive argument method closely resembles the medieval Arabic method, and indeed, derives from it. He makes a powerful case for the transmission of the institution of the *madrasa*, or Islamic college, and much of the structure of a higher education institution, to Western Europe at the same time as the appearance of the recursive argument method, arguing that because the method was used in oral disputation in very similar ways in both the *madrasa* and the college, the two were connected.[39]

Makdisi claims that the medieval legal scholar Ibn ʿAqīl's arguments in part of his summa contain a simple form of recursive argument method. He says (1981: 255) that Ibn ʿAqīl's argument structure "is reducible to the basic schema, namely: (1) thesis and counter-thesis; (2) arguments for the thesis; (3) objections to the arguments; (4) replies to the objections; (5) pseudo-arguments for the counter-thesis; and (6) replies in refutation of these pseudo-arguments." This may be diagrammed for clarity as follows:

I. Thesis (*Pro*)
II. Counterthesis (argument *Contra* I)
III. Arguments for the Thesis (arguments *Pro* I)
IV. Arguments against the Thesis in III (arguments *Contra* I)
V. Arguments against the arguments in IV (arguments *Pro* I)
VI. Arguments against the argument in V (arguments *Contra* I)
VII. Arguments against the arguments in VI (arguments *Pro* I)

Thus, although Makdisi claims that the arguments of Ibn ʿAqīl and Thomas Aquinas are essentially the same, with the exception of the location of the AUTHOR'S VIEW, in fact the arguments are simply an ABABABA dialogue type of argument structure, in which A represents the first argument or further arguments in support of it and B represents the arguments against A. It is therefore not a recursive argument such as that used by Avicenna and the Medieval Latin authors of the thirteenth century on. Makdisi does not cite any examples of an actual recursive argument in Arabic works and seems to be unaware of the fundamental difference between the two argument structures.[40]

[39] Makdisi (1981, 1990); cf. Grant (1996: 40–41) for a similar argument.

[40] He does not give an example of Ibn ʿAqīl's arguments. However, he has done a great service to scholarship in producing a critical edition of Ibn ʿAqīl's huge summa.

Significantly, as Makdisi notes, some have argued that the source of the European recursive argument method goes back as far as the *Sentences* of late classical scholars such as Isidore of Seville (d. 636) or even earlier. "Long before Abelard, a disciple of the Sophist Protagoras had compiled 'a dull catalogue of mutually conflicting opinions in about the year 400 (B.C.)'. Protagoras 'is said to have been the first person to teach that it is possible to argue for or against any proposition whatsoever'."[41] However, because there is absolutely no question here of a method of "reconciling pros and cons," not to speak of the recursive method, Makdisi's argument actually supports the view—long accepted by scholars who have worked closely on the history of the recursive argument method—that it has no classical or early medieval ancestor.

Makdisi repeatedly refers to the strange idea that the recursive argument method is a procedure for "reconciling" various "contradictory" views. Following many earlier scholars and their idea that the *Sic et non* is one of the earliest precursors of the Latin recursive argument method (he believes the legal scholars to be the absolute earliest), he says, "In this work Abelard (d. 1142) cites a series of affirmative sentences, matching them with a series of negative sentences, all by Fathers of the Church. The Prologue gives explicit instructions on the method of reconciling these pros and cons, but the author does not apply these rules, and he makes no attempt to reconcile these apparently contradictory opinions."[42] As for the legal scholar he believes to be the earliest to use the method, he says, "Ivo of Chartres (d. 1116) is another canon lawyer who made use of the sic et non method before Abelard. In the Prologue of his *Decretum* he cites the rules for reconciling conflicting texts."[43]

Many scholars have also claimed that besides Gratian and Abelard, additionally or alternatively Anselm of Canterbury, Peter Lombard, Robert of Melun, or some other scholar who wrote before the mid-twelfth century first uses the Latin recursive argument method, which they claim is an organic growth out of the local Western European theological tradition. However, these claims are always based on the *content* of the texts rather than the *overt, explicit argument structure*—and indeed, some believe that the defining characteristic of the "scholastic method" is *scholasticism*—a term referring to the content, or supposed general

[41] Makdisi (1981: 338 n. 82).
[42] Makdisi (1981: 246).
[43] Makdisi (1981: 247).

philosophical approach, of the texts.[44] Nevertheless, they generally also claim that the unusual overt structure of the recursive argument method, or at least its origin, is to be found in such works, especially the *Sic et non* of Abelard and canon law texts such as the *Decretum* of Gratian.[45] None of this is correct, as examination of these texts immediately reveals. Consider, first, the overt structure of an item from the *Decretum* of Gratian (fl. mid-twelfth century), a work on canon law presenting different views on specific legal questions.

### *Example 2.2*

**GRATIAN**
***Decretum***[46]
DISTINCTION NINETEEN

PART 1.

There is a question whether decretal letters have authoritative force, because they do not appear in the corpus of canons.
*Concerning these, Pope Nicholas writes to the archbishops and bishops of Gaul:*
C.1. *Decretal letters have authoritative force.*
§1. Works of other writers were approved or rejected by decree of the Roman pontiffs, and so today the Apostolic See . . .
§2. But those more defiant than obedient will reply that among the canons there is a capitulum of Pope St. Innocent . . .
§3. From the preceding, with the help of divine grace, we have shown that there is no difference between . . .
*Also, Pope Agatho to all the bishops:*

[44] E.g., Wulf (1956: 37 et seq.). In fact, much research on the topic of the scholastic method has focused on the *content* of some of the most notable books that use it (especially works on metaphysics and theology), or on logic, the rules of oral debate, the technical terms for points of defeat, and so on. Although all of these topics have classical or early medieval precedents, none of them have any essential connection with the overt, explicit structure of the recursive argument method, which has no such precedents in Greek or Latin Europe.

[45] On these proposals in the earlier literature, see Makdisi (1981: 245–253), who agrees with them, despite his own argument that the scholastic method was borrowed from the Islamic world along with the college.

[46] English: Thompson, Gordley, Christensen 1979: 76–84, q.v. for the complete text. The selections given here are quoted verbatim, including all organizational marks or numbers. The only thing I have done is to modify the typography according to the style of the present book, and to include the marginal glosses in brackets when helpful for understanding the quoted portions of the text.

C.2. *All sanctions of the Apostolic See are to be observed inviolably.*
All of the sanctions of the Apostolic See are to be received as confirmed by the voice of St. Peter himself.
*Also, from the capitula of the emperor Charles:*
C.3. *A yoke that has been imposed by the Holy See is to be endured even if it seems insupportable.*
§1. In memory of the blessed Apostle Peter, let us honor the holy Roman and Apostolic See so that . . .
§2. Now, if someone, whether priest or deacon, in order to stir up any unseemly disturbance or to undermine our ministry . . .
*Also, Pope Steven V:*
C.4. *Whatever the Roman Church decrees or ordains is to be observed by all.*
To be sure, because the holy Roman Church, which Christ appointed to rule over us, is set up as a mirror and example, whatever she decrees and ordains is to be perpetually and inviolately observed by all.
*Also, Gregory IV:*
C.5. *He who refuses to obey apostolic precepts is unsuitable for pontifical office.*
It is wrong that anyone try to transgress or be able to transgress the precepts of the Apostolic See . . .

Part 2.

Therefore, let anyone who would contradict apostolic decrees be cast down to his sorrow and ruin, and let him no longer have a place among the priests. . . .
*Also, Augustine in* On Christian Doctrine, *II, viii*:
C.6. *Decretal letters are to be reckoned among the canonical writings.*
§1. In regard to canonical writings, let the careful student of the divine Scriptures follow the authority . . .
§2. Accordingly, he will observe this rule concerning the canonical writings: he will prefer those that . . .
*Also, Pope Leo I to the bishops of Vienne, in* Letter *lxxxvii*:
C.7. *Let one who abandons solidarity with Peter know himself to be deprived of the divine ministry.*
Our Lord Jesus Christ, the savior of the human race, determined that the truth contained earlier in the proclamation of the Law . . .
*This* [Gloss: *This*—namely, that decretal letters have the same authority as canons.], *however, is to be understood only of those sanctions or decretal letters in which nothing is found contrary to the decrees of earlier fathers or evangelical precepts . . .*

*Anastasius II to Anastasius Augustus,* Letter *i, 7–8:*

C.8. *No share in the offense is attached to one who has been ordained by previously condemned heretics.*

Your Serenity's heart knows that according to the most sacred custom of the Catholic Church none of those whom Acacius [Gloss: *Acacius*—a heretic and excommunicate . . . ] baptized or ordained . . .

*Because he*[47] *gave this rescript illicitly, uncanonically, and contrary to the decrees of his predecessors and successors . . . he has been repudiated by the Roman Church and it is read that he was also smitten by God:*

C.9. *Anastasius, reproved by God, was smitten by divine command.*

Anastasius II, a Roman by birth, lived in the time of King Theodoric. At that time, many clerics and priests . . .

*Thereupon, it was decreed concerning Bishop Maximus, at the Synod of Constantinople under Pope Damasus, c. 6:*

C.10. *Let everything done by or in conjunction with undisciplined prelates be revoked as void.*

Because of the report of his complete lack of discipline, which was revealed at Constantinople, it is decreed that Maximus is not to be considered to have been a bishop at any time . . .

Gratian's method obviously is to state a topic and then list different legal arguments after it. The format of his text is similar to the first half of a recursive argument, but because it has no recursive sequential disputation of the subarguments it is not the recursive argument method.

Similarly, the *Sic et non* of Peter Abelard (d. 1142), a compendium of contradictory passages in Christian Scripture and theological works, gives a topic and a list of subarguments, but again, without recursion or indeed any disputation of the subarguments, as shown in Example 2.3.

### *Example 2.3*

**Peter Abelard**
***Sic et non***[48]
Question 55

[Topic]

That Eve alone, not Adam, was the seducer, and the contrary

[47] I.e., Pope Anastasius II.

[48] Latin: Boyer and McKeon (1976–1977: 231–232), q.v. for the full Latin text and precise identifications of the works cited by Abelard. Incredibly, there seems as yet to be no translation of Abelard's *Sic et non* into English or any other modern language.

[ARGUMENTS]

[1] Augustine, in *Super Genesim ad litteram*, Book XI: The apostle says: Adam is not seduced, the woman is . . .

[2] Ambrosius, in *Exameron de die*, V: Adam is deceived by Eve, not Eve by Adam . . .

[3] Paschasius, *De corpore et sanguine Domine*, Chapter XII: Divine wisdom . . .

[4] From the *Sermons of Chrysostom*, in XL: Then the consenting woman . . .

[5] Augustine, *De singularitate clericorum*: It is wondered, if Adam is seduced through Eve . . .

Makdisi remarks, "Grabmann considers the so-called influence of Abelard's *Sic et Non* on the *Decretum* of Gratian as 'certainly very much overrated'." He also notes recent work "showing Abelard's influence on the canonistic movement of the twelfth century, especially with respect to Gratian's successors." Like Kantorowicz and others, Makdisi argues that the recursive argument method is the product of the canon law tradition, which he compares to the equivalent tradition in the Islamic world. There is a real possibility that the recursive argument method was indeed transmitted orally via public disputation in Islamic legal colleges, as Makdisi argues at some length, but unfortunately, he never presents actual textual evidence from Islamic and medieval European jurisprudence to support his theory. It seems that so far no one else has yet attempted to do an in-depth study of the Islamic legal literature in order to test Makdisi's theory and compare its argument methods with those of the Medieval Latin legal scholars. Until that is done, and unless valid examples of the recursive argument are then found in both literatures, little more can be said.[49]

The legal scholar Gratian's *Decretum*, and the theologian Abelard's *Sic et Non*, the two most widely cited putative sources of the recursive argument method, contain only lists of positions, with no recursive argumentation, no establishment of the author's view, and no actual suggestion in them that anyone complete these supposedly incomplete works by using anything like the recursive argument method. Their "argument" structure, such as it is, is not even remotely related to the overt structure of the recursive argument used by the great Latin scholastics from the early

[49] Makdisi (1981: 339 nn. 95–96).

thirteenth century on. Neither Gratian nor Abelard, nor any of the other works of that type, can therefore be the source of the method.

In short, there are many examples of the earlier medieval "questions" or "sentences" type of literary format, such as that used by Gratian and Abelard (see Examples 2.2 and 2.3), which has been argued by many to be the direct ancestor of the recursive argument method, but none actually contain recursive arguments and therefore they cannot be the source of the method.

In the early thirteenth century, several older works of the *quaestiones* ('questions') or *sententiae* ('sentences, opinions') type, such as the *Questiones* of Peter the Chanter and the *Sententiae* of Peter Lombard, were revised or completely rewritten according to the latest method by their students or disciples, and presented as their own works.[50] One such revisor was Robert of Curzon, whose summa has long been thought to be the earliest original Latin work to use the recursive method.[51] The *Sententiarum libri quinque* 'Five Books of Sentences' of his near contemporary, Peter of Poitiers (whose dates are uncertain), contains some examples of the recursive argument method, but its chronological precociousness is highly suspect. It seems clear that Peter of Poitiers or one of his students later modernized the work according to the new recursive argument method.[52] These texts therefore do not represent a straight-line development out of the earlier nonrecursive Latin methods, but instead a revision of older texts to accomodate them to an entirely new method while retaining much of the traditional appearance of the old methods.

The overt, explicit, formal structure of the recursive argument is its most crucial feature.[53] It is not quite true that "the medium is the message" in recursive method books, but because they typically consist exclusively of lists of recursive arguments, each of which contains many contrasting views on the same problem, they clearly did encourage scepticism and speculation by the authors. In that respect, therefore, it is true that the form of the recursive argument did have a significant indirect impact on the content of works written according to it.[54] Nevertheless,

[50] Baldwin (1970: 100–101).

[51] See chapter 6.

[52] See appendix B, in which the chronological problems and many other problems involving Peter of Poitiers are discussed.

[53] Cf. Grant (2007: 182–183).

[54] See Grant (2007: 325ff.) on the importance of the recursive argument method in encouraging a "probing and poking around" approach, a kind of curiosity about the world both as perceived and purely as imagined, and on the impact of the method on the development of scientific thought in Europe in general.

it must be stressed that the specific overt structure, per se, of a recursive argument is not directly or even implicitly connected, structurally or semantically, to its specific overt content or to the implicit logical structure of the internal content. In other words, in a recursive argument method, the *way* it is said has essentially nothing to do with *what* is said. It does, however, have a great deal to do with the general way the content is approached and understood, as became clear with the adoption of the method for scientific works from the mid-thirteenth century on. In this respect, the dictum "the medium is the message" reflects the very long life of the method, and what might be called its afterlife as the "ideal" literary scientific method of modern times.[55]

[55] See chapter 8.

— Chapter Three —

# From College and *Universitas* to University

Science apparently requires a permanent, independent, self-supporting institution devoted to advanced learning, in order to ensure the continuity and growth of knowledge from generation to generation and provide a tradition or "normal science paradigm" against which revolutionary scientists young and old can struggle.[1] It is uncontested that the first college in Europe was established in Paris in 1180. It is also established that the early *universitas* of the period was actually a trade guild; it was not even remotely like a university in the sense of the word *university* or its equivalents in modern European languages for the past half millennium. So, where did the college and later "true" university come from?

The *madrasa*, the medieval Islamic college, appeared in Central Asia at least two centuries before the first college founded in Western Europe. This is the traditional, accepted view among Islamicists. However, scholarly explanations for this fact do not agree. It is thus necessary to discuss for a moment the view that now dominates the Islamicist literature.

The standard reference today is a long *Encyclopaedia of Islam* article on the *madrasa* by Pedersen, Makdisi, and others, which is slanted heavily toward the view that the *madrasa* grew organically out of earlier, purely Arab Islamic institutions, particularly the mosque-school, and argues that the *madrasa* was only a minor legal modification of it:

> One should not attach undue importance to the fact that the *madrasa* developed especially in the eastern lands of the caliphate, in Iraq, Persia and Transoxiania; this does not imply a cultural swing away from Arab Baghdad towards Persian Khurasan, especially towards Nishapur, which would be a misreading of cultural history due to anachronistic nationalist sentiment. Since there was no change in the curriculum, or in the teaching staff or

[1] See chapter 7.

> students, and since the final product of the two types of colleges [i.e., *madrasa* and mosque-school] was exactly the same, that is, the *muftī-faqīh*, the reason for the change in institutional typology must be sought elsewhere and may be found in the legal status of the two institutions concerned, the *masjid* [mosque] and the *madrasa*.[2]

This "internal" argument, a clone of the Europeanist argument about the origin of the college in the cathedral-school, claims that the *madrasa* originated in schools connected to the mosque (a purely Arab Islamic architectural form and Arab Islamic institution), and therefore the *madrasa* did *not* first appear in Central Asia. But this view is explicitly refuted by the archaeological, architectural, and historical evidence. The *madrasa* was an unprecedented innovation in the Islamic world. The earliest attested *madrasas*, and *madrasas* in general, were long found only in Central Asia. The same article points out the sharp physical distinction between mosque-schools and *madrasas* known to architectural historians for a century, if not longer. Unlike mosque-schools, *madrasas* "were especially arranged for study and the maintenance of students."[3] Makdisi emphasizes that the *madrasa* was almost exclusively devoted to legal education. This is to a large extent correct—if not before, then certainly after the standardization of the institution under the Seljuks and its spread across their vast empire. Although some of the best-known literature produced by some of the most notable *madrasa* teachers, such as al-Ghazālī, is anything but legal in its subject matter—ranging from theology and metaphysics to Sufism and belles lettres (and in many of the earliest well-known *madrasas*, those in the Central Asian city of Nishapur, still other subjects were taught)[4]—the fact remains that other than al-Ghazālī, none of the great natural philosophers of Classical Arabic civilization studied or taught in *madrasas*. Indeed, the most famous scholarly work written by al-Ghazālī himself is actually an attack on science, scientific thought, and scientists. It encouraged a sociopolitical turn against natural philosophy that eventually resulted in the effective death

[2] Pedersen et al. (1986). The article's Arabic transcriptions have been normalized to accord with standard international scholarly practice, except for well-known toponyms or other words, which follow traditional English spellings.

[3] Pedersen et al. (1986).

[4] Bulliet (1972). Another problem with the *E.I.*$_2$ article is that it often refers to 'schools' without specifying which Arabic term the word is supposed to translate. This seems to have been intended to support their argument about the putative Arab Islamic origin of the *madrasa* and its supposed unimportance within Islamic education.

of science in the Islamic world.[5] The great importance of the *madrasa* for science is ultimately what happened to it in Europe.

It must be stressed that the college and the *universitas* originally were completely unrelated institutions. The *universitas*, an incorporated scholars' guild (something like a modern trade union), was a native European development. The college was radically different in every way. Simple comparison of the medieval Islamic *madrasa* with the medieval European college[6] has already shown that these two share their fundamental institutional features.[7]

The actual charter of the first college known to have been founded in Western Europe,[8] the Collège des Dix-huit, which was established in Paris in 1180, states that it was founded by Jocius of London, a wealthy English merchant[9] who had just returned "from Jerusalem."[10] By the period of Jocius's visit to the Near East, *madrasas* were very common there.[11] Like the *madrasa*, the college is an all-inclusive academic institution with a permanent endowment[12] recognized by the government. The endowment, in both the Islamic and Western European traditions, covered the expenses of the physical property and living support for the scholars—the students and their teacher or teachers—all of whom lived together in the same structure. Based on the brief description in the founding charter and what is known about other early colleges from the following decades, including the Sorbonne, the college founded by Jocius is identical in all particulars to the typical *madrasa* then widespread in Syria and its vicinity.[13] They

[5] See chapter 7.

[6] It was not called a *collegium* 'college' at first. Similarly, as noted above, it is firmly established that the *universitas*, despite its name—which is usually wrongly rendered as *university* in Modern English—was not a university at all in anything like our Modern English sense of the word; it was a scholars' guild (Grant 1996, Rashdall 1936). Although the early colleges are little known, such that the details of their operation are not clear until well into the thirteenth century, the most distinctive features of the two institutions are evident from the outset. They continued to be distinctive until they began to merge, fairly early, and eventually gave birth to the modern university. The important thing in these cases is once again not the names but the design and purpose of the references of the names. See the comments on terminology by Grant (2007: 319ff.), and see further below.

[7] Makdisi (1981).

[8] Due to the adherence of the Andalusian Muslims to the Maliki school of jurisprudence, there seem to have been no *madrasas* in Spain until after the college had already appeared in France (Makdisi 1981).

[9] Ferruolo (1985: 284); cf. Rashdall (1936: 502–504).

[10] Denifle (1899, 1: 49). See appendix C for the full text of the charter.

[11] Makdisi (1981: 228).

[12] In the Islamic world, it is a type of *waqf*. Surviving *waqf* documents specifically for *madrasas* are said to be rare, late, and largely unstudied.

[13] Hillenbrand (1986: 1138–1139). The endowment, organization, and other institutional features of the early Buddhist-Islamic college in Central Asia also remain little studied; see, however, Van Bladel (2010).

were endowed institutions, generally quite small, which housed a small number of students, typically less than two dozen: exactly like the Collège des Dix-huit and most of the other early colleges. Because Jerusalem is located inland, Jocius had necessarily spent time in the Islamic Near East—undoubtedly in Syria, which was one of the main destinations of merchants and pilgrims alike. There he must have encountered the local small type of *madrasa* on which he modeled the identical institution he founded in Paris, Europe's first college. The Near Eastern origin of the Western European college could hardly be clearer.

But there is much more to this story. It has actually been known for almost a century, though little recognized by most scholars, that the medieval Central Asian Islamic college, the *madrasa*, is an Islamicized form of the earlier Central Asian Buddhist college, the *vihāra*. The two are virtually identical in form, function, teaching program, and legal status. The identity of the unique architectural form of the Central Asian *vihāra* and *madrasa* has also been established by archaeologists,[14] whose discoveries have confirmed the much earlier arguments of Barthold.[15] The typical Central Asian *vihāra* had a square or rectangular plan, which is replicated in the *madrasa*.[16] In addition to the combination of functions and endowment, the discovery and excavation of a pre-Islamic example of the *vihāra* found at Adzhina-Tepa in Tajikistan make it absolutely certain that the typical rectangular four-*īwān*-style[17] *madrasa* is simply an Islamicized version of the identical four-*īwān*-style *vihāra*.[18] Of the

[14] Litvinskij and Zeimal' (1971). Note that the English translation of this book is nearly gibberish, but the original Russian edition (cited here) has a long summary in normal English.

[15] Barthold (1964: 30), originally published in 1918; Litvinskij (1985); Pedersen et al. (1986). The "Arabocentric" arguments in the $E.I._2$ article on the *madrasa* have been resoundingly disproved by archaeology and architectural history, above all the discovery and full excavation of Adzhina Tepa. The issue for science now is to move onward with further study of the clear connection between the *vihāra* and the *madrasa*.

[16] The remarkable circular plan of the Nawbahār (Sanskrit *Nava Vihāra* 'New College') of Balkh, the home of the famous Barmakid (Barmecide) family, was completely unprecedented. It is an artifact of the structure's original design and function as a Sasanid imperial palace (the circular plan of which was based on the Parthian variant of the Central Eurasian circular *ordo* 'royal encampment' plan). For descriptions of the structure, see Beckwith (1984b) and Van Bladel (2010); cf. chapter 5, note 24. The Nawbahār was the most famous and intellectually important Buddhist *vihāra* not only in Balkh (anciently Bactra, the most important city of Bactria, or Tokhāristān, a thoroughly Buddhist country stretching from west of the city east into what is now Tajikistan), but in pre-Islamic western Central Asia as a whole. It evidently continued to function for a century or more after the Arab conquest, as did many other *vihāras*. See Beckwith (2009: 123, 147, 394 n. 28) for emendations to the description of the Nawbahār in Beckwith (1984b).

[17] Some later *madrasas* have only one or two *īwāns*, or even none. Perhaps the most beautiful surviving examples of the traditional Central Asian four-*īwān madrasa* are Renaissance-period structures in Isfahan.

[18] Litvinskij and Zeimal' (1971), Hillenbrand (1986: 1136).

many *vihāras* that once existed in the area of Khuttal (in what is now southern Tajikistan), archaeologists have excavated the remains of several, including the one at Adzhina Tepa, which was destroyed during the mid-eighth-century wars in the region.[19] It already has what became the stereotypical Central Asian and Iranian *madrasa* plan: a quadrangle with one or two floors of arcaded hallways opening onto a central courtyard. Behind the arcades are cells for the resident scholars. In the center of each side is a large half-domed space—an *īwān*—the open side of which faces the courtyard. Some *madrasas* have one side physically attached to a mosque or a shrine or both, which means fewer arcades and cells for scholars.[20]

The earliest examples of *vihāras* built according to the plan of Adzhina Tepa (though without *īwāns*) have been found in the ruins of the great city of Taxila (Takṣaśīla), dating to the period of the Kushan Empire (ca. 50 BC–AD 225). This empire was a powerful Central Asian state, founded in Bactria,[21] which included in its territory Gandhāra (the southeasternmost region of Central Asia), the great city of which was Taxila (located very near to what is now Islamabad-Rawalpindi in northern Pakistan). No pre-Kushan-period *vihāras* have been found,[22] and the plan of the *vihāra* is strikingly different from that of the *saṅghārāma*, the typical earlier, strictly Indian, Buddhist monastic design.[23] The *vihāra* design is thus a specifically Central Asian innovation developed under the Kushans and spread by them.

The *vihāra* also had a special legal status. It was supported by a tax-free pious endowment (known in Arabic as a *waqf*). Hui Ch'ao (fl. ca. 726) comments, "Whenever a *vihāra* is built, a village and its inhabitants are immediately donated as an offering to the Three Jewels.[24] Building a *vihāra* without making any donation of a village and its folk is not done. . . . Since there are no slaves [in India], it is necessary to donate villages and their inhabitants."[25] The best-known Central Asian *vihāra*, the Nawbahār

[19] Litvinskij and Zeimal' (1971).

[20] The same variant is found in European cloister plans, in some of which (e.g., Magdalen College, Oxford) one side of the quadrangle is occupied by a church.

[21] Beckwith (2009: 85).

[22] Dutt (1962: 211ff.).

[23] Dutt (1962: 62ff.).

[24] I.e., to Buddhism.

[25] Van Bladel (2010: 67), substituting *vihāra* for his *monastery*. Hui Ch'ao's remark makes it clear that what peripheral peoples often called 'slaves' in Central Eurasia were virtually never 'slaves' in the modern sense. Moreover, "The monastic codes (*vinayas*) specified that land properties were held by the monastic establishment in perpetuity" (Van Bladel 2010: 67 n. 106). The *vinayas* of all the major Buddhist traditions are thought to go back to India.

(*Nava Vihāra* 'New Vihāra') of Balkh, is also the best-described: "They had granted the Barmak seven *farsakhs* in diameter[26] of the lands around the Nawbahār . . . and a farm district called Zuwān in Tokharistan, eight *farsakhs* by four *farsakhs*.[27] . . . The house had many endowments of property (*wuqūf*) and great estates."[28]

The earliest historically mentioned *madrasa* noted so far under that name in the scholarly literature[29] is the school endowed by Abū Ḥāṭim al-Bustī (890–965)[30] in his native town, Bust. It had apartments and scholarships for his students and foreign students, and a library.[31] This is exactly like the Buddhist *vihāra*. The *vihāra-madrasa* must have already spread widely by the middle of the tenth century.[32] By 1025–1026 there already were twenty fully endowed *madrasas* in Khuttal (then part of Tokharistan, now part of Tajikistan), the very same province where once the former *vihāra* of Adzhina Tepa and the hundreds of *vihāras* noted by Hui Ch'ao when he passed through in ca. 726 AD were operational.[33] Moreover, in one of the major metropolises of Central Asia, Nishapur, "no less than 38 madrasas *predating* the great Niẓāmiyya of that city (founded ca. 450/1058) are recorded."[34] It may be assumed from similarities in the teaching methods and in the main philosophical interests—theology

[26] Van Bladel (2010: 65 n. 94): "Perhaps rather 'seven square *farsakhs*'."

[27] Van Bladel (2010: 65 n. 95) notes that the text of Ibn al-Faqīh edited by De Goeje contains this passage, which is not found in the Mashhad manuscript. I have removed his editorial brackets.

[28] Text translated by Van Bladel (2010: 64–65) from the Mashhad manuscript of Ibn al-Faqīh (van Bladel 2010: 62–63); *wuqūf* is a plural form of *waqf*.

[29] One reason that no earlier references to the institution have been found could be that the name *madrasa* replaced an earlier name for it. If so, was it perhaps a loan form of *vihāra* (or another equivalent)? Another reason is that after the collapse of central power in the ninth century, Arabic historical knowledge of Central Asia drastically declined. And yet another reason is Islamicists' apparent lack of interest in the history of the *madrasa*.

[30] Wüstenfeld (1891: 158–165). Wüstenfeld (1891: 163) evidently translates *madrasa* into German as *hohe Schule*, i.e., 'advanced school'. He gives more information about the school, based on a report of al-Khaṭīb al-Baghdādī, but he unfortunately does not give a reference to the exact location of the report in that author's voluminous works.

[31] Pedersen and Makdisi (1986: 1126), citing Wüstenfeld (1891: 163). Bust (also Bost), is now Lashkar Gāh, in Sijistān (southwestern Afghanistan).

[32] *Pace* the "nativist" views of Pedersen and Makdisi (1986). Cf. Hillenbrand (1986: 1136), who comments, "the earliest *madrasas* recorded are those of eastern Iran in the early 4th/10th century." He also adds the understatement that the theory of derivation of the *madrasa* from the Khurasani house "cannot be more than speculation." Note that "eastern Iran" here, as often in Islamicist works, actually refers to Central Asia, not "Iran" (Persia), though the northeastern part of northern Iran, Khurasan, was traditionally a part of Central Asia except during periods of Persian imperial expansion.

[33] Fuchs (1938: 452); cf. Litvinskij (1985).

[34] Hillenbrand (1986: 1136), emphasis added. Bulliet (1972: 249–255) gives historical and biographical details about these *madrasas*.

and religious law, both comprised in the Buddhist context by the word *dharma*[35]—that the Muslims in Central Asia had continued the *vihāra* pattern intellectually, too.

Some might ask at this point how any of this could possibly relate to the European university, a large, wonderful institution that many believe is purely European in origin and developed long before the small, unimpressive college. However, this traditional view is based on projection backward of the later situation and thereby distorting and obscuring the historical development of the European higher education system.

First of all, it is clear that the earliest three "universities"—the *universitas* guilds of Bologna, Paris, and Oxford—appear at approximately the same time in history, the late twelfth or early thirteenth century; regardless of scholarly tradition, none has been demonstrated with any certainty to be significantly earlier than the others.[36] However, it cannot be overemphasized that although the Latin word *universitas* is usually "translated" as *university*,[37] the early *universitas* was totally unlike a *university* in its purpose, structure, and functions. The word *universitas* originally meant an "incorporated" guild of any kind. Although the word *universitas* later acquired the meaning of a 'guild of scholars' specifically,[38] even then it still was strictly a guild; it was nothing like a university: "The *universitas* was in its origin a voluntary association of individual masters rather than a single educational institution conducted by an organized staff. The *universitas* prescribed the studies which were to lead to the master's chair; but it did not attempt to interfere with the discipline of the scholars. In a sense all scholars were

[35] The normal translation of *dharma* in Chinese is *fa* 法 'law'. This is significant because Buddhism first came to China from Central Asia.

[36] See Grant (1996: 36–37); cf. Grant (2007). See also Rashdall (1936), a careful reading of which, omitting his many speculations about supposed gradual internal development of this or that new feature, also makes this clear. It appears that the "University of Paris"—i.e., the *universitas* guild of the masters in Paris—cannot be dated firmly *as a corporate body* earlier than the papal statute of 1215 issued by Robert of Curzon, though it evidently existed by 1210, and existed in name by 1212 (Rashdall 1936: 308–309). Whether or not the tradition by which Bologna is the oldest *universitas* of scholars, and Oxford the youngest, is correct, they nevertheless still did not become universities, in the sense of our word *universities* (i.e., rather than *universitas* guilds), until they began merging with the college.

[37] More precisely, the Latin word *universitas* is converted to English *university* or another modern language equivalent, as if the two meant the same thing. This is a pernicious practice that pervades European studies in general.

[38] Grant (1996: 34ff.), Rashdall (1936: 286ff.). The *universitas* guilds are often called by Rashdall and other modern scholars (e.g., Makdisi 1981: 224) 'corporations', but this term too did not mean then quite what it means today and is best avoided.

regarded as members, though not as governing members, of the 'Universitas of Masters and Scholars'."[39]

In short, the scholars' *universitas*, however influential we might think it was, long remained a guild, pure and simple: "At the outset it [the term *universitas*] was applied to a single group that formed a legally recognized self-governing association. Thus a faculty of arts was a 'university'[40] as was any faculty of medicine or faculty of theology. The masters and students of the arts faculty formed their own legal corporation, or university, as did the teachers and students of the medical faculty, and so on."[41]

The early *universitas* guilds of scholars did not own buildings or other physical property, they were not supported by permanent financial arrangements such as pious foundations, and they did not have much of anything else that we think marks an institution of higher education as such. The only significant thing the early *universitas* guilds did have that we would recognize as related to the function of a university was the right to bestow an advanced degree—the 'license to teach'[42]—and this has been shown to be a borrowing from the earlier attested *ijāza li-'l-tadrīs* 'license to teach' of medieval Islamic culture.[43]

> The term that was initially employed, and was in common use by the middle of the thirteenth century, to encompass all of these individual, disparate universities, or university associations,[44] was *studium generale*. Every master and student was a member not only of his own individual university, or corporation,[45] but also of the *studium generale*. . . . The term was usually assigned to schools that either were sufficiently prestigious, such as the customary universities of Paris, Oxford, and Bologna, or were large enough

[39] Rashdall (1936: 521); I have substituted the historically correct term *universitas* for Rashdall's misleading modern term *university* in every instance in this quotation.

[40] I.e., a *universitas* guild, not a 'university' in our sense.

[41] Grant (1996: 35).

[42] Makdisi (1990: 26–27). Other than the right to bestow degrees, the main right or privilege of the *universitas* guild was rent control for students living in a *hospicium*, granted in the statute of 1215 promulgated by Robert of Curzon (Rashdall 1936: 498–499). The scholars (masters and students) who belonged to the *universitas* were actively protected by the royal government on the one hand, and by the Church on the other, which gave them clerical status, meaning financial support from the church and better legal treatment if they were arrested by the authorities (Grant 1996: 36), though this was not significantly different from the rights enjoyed by many other nonscholastic clerical groups, notably the religious orders.

[43] See further below in this chapter.

[44] I.e., *universitas* guilds.

[45] This again refers to the *universitas* guild.

to include at least three of the four traditional faculties (arts, theology, law, and medicine), or were both.[46]

But by the mid-thirteenth century, when the term *studium generale* came into general use, the *college* had already spread everywhere too. Its influence on the *universitas*, and vice versa, was such that a new institution developed out of both, namely, the university in the modern sense. The college is considered by Verger to have been the most dynamic new feature in Western European higher education.[47] The term *university* replaced *studium generale* by the end of the Middle Ages,[48] marking the merger of the *universitas*, the *studium generale*, and the college into the early modern college-university.

Soon after the first college, the Collège des Dix-huit, was established in Paris in 1180,[49] more colleges were founded, about which little is known,[50] but based on later information, it seems that many were intended for students from one of the four "nations" into which the arts faculty of the university was divided. The Sorbonne, founded in or around 1257 by Robert de Sorbon, was a college specifically for sixteen advanced students of theology, four from each nation,[51] who already had the degree of Master of Arts.[52] The colleges were endowed institutions with their

[46] Grant (1996: 35).

[47] "La mutation la plus décisive est l'émergence du collège," among the "phénomènes majeurs bouleversent les Universités françaises à l'époque moderne" (Verger 1986: 141, cf. 38–39, 80); cf. his discussion of the Sorbonne (Verger 1995: 64ff.). However, he assumes the local, internal development of the college within Europe.

[48] Grant (1996: 35–36).

[49] By 1231 the college had its own house near the church of St. Christopher. The college survived until 1789 (Rashdall 1936: 502).

[50] Rashdall (1936: 501ff.). On the earliest colleges he notes, "It is not clear whether they [the students] originally lived under the supervision of a master, though they certainly did at a later date" (Rashdall 1936: 505). His idea that the colleges were establishments for small boys seems to have been created out of whole cloth, based on the extremely laconic accounts of the earliest colleges, which usually mention only that they were for the benefit of "poor clerks," who in slightly later documents are specified as having been scholars. The earliest example of a college founded for "young and poor boys" (but apparently still not "small" boys) that he cites is actually dated 1339, more than a century and a half after the foundation of the Collège des Dix-huit (Rashdall 1936: 510–511).

[51] The Sorbonne was exceptional in several respects. This particular detail apparently reflects the division of the arts faculty of the University of Paris into four "nations," based on the geographical origin of their members: the French, Picard, Norman, and English or Anglo-German. The proctors of the four nations elected the rector of the university (Grant 1996: 37; cf. Verger 1986: 38).

[52] Rashdall (1936: 507), who notes that the all-nation inclusiveness of the Sorbonne was "an unusual feature in Parisian college constitutions, and perhaps laid the foundations of the future greatness of the college." The number of scholars was soon increased, by additional benefactions, to thirty-six.

own physical property secured by their endowments and recognized and protected by the government, though the Sorbonne, unlike the other colleges, was not governed by its own master.[53] The colleges also had their own libraries[54] (as the *madrasas* and still earlier the *vihāras* did). They thus had the essential features of a modern university (or American 'college') *before* the "University" of Paris had become a university in the modern sense by gradually merging with its colleges, partly by forcibly taking them over.[55] The colleges of Paris thus literally laid the foundations for the actual University of Paris.[56]

The subsequently founded universities of Europe mostly[57] followed the early Parisian model at first, with a *universitas* guild of masters plus a number of colleges. The best-known example is Oxford, which followed the early Parisian model very closely. However, its colleges took a slightly different historical direction almost from the moment they were established. University College, the first at Oxford, was founded in 1249 with a bequest made by William of Durham. It followed the early Parisian model. The next college to appear there was Merton College, founded in 1264. It was the second Oxford college, but the first to be incorporated, and was followed by many others.[58] The powerful combination of the endowed college and the corporate guild of scholars ensured that the early English universities would keep their colleges, which have retained their influence down to the present in those institutions.

As for the architectural form of the European college, it is a fact that the 'cloister' of the early English college happens to be essentially identical physically to the *madrasa-vihāra*, and this design seems to have been

[53] Rashdall (1936: 508ff., 516); Verger (1986: 39).

[54] Rashdall (1936: 516–517).

[55] Rashdall (1936: 522ff.), who notes the marked difference from the situation in the English system, where this did not happen.

[56] Verger (1986: 141ff.).

[57] According to Makdisi (1981: 237), the earliest public university, founded by the proclamation of King Alfonso VIII of Castile in 1208–1209 and funded by the royal government, was the University of Palencia, where the masters were invited from abroad to teach for salaries. Makdisi suggests that the Spanish "radical departure from the original idea of the university was inspired by . . . the experience of Western Islam, where a peculiarity of the predominant Maliki law *waqf* discouraged all but sovereigns to found colleges" (Makdisi 1981: 238, italics corrected). However, one must note that whatever was founded by Alfonso VIII, it could not have been an actual university (in the sense of our word *university*), or a college of any kind, but undoubtedly a *universitas*, so it could hardly have been inspired, institutionally, by the *madrasa*. Even so, its history is certainly worth further study.

[58] Makdisi (1981: 227–228); Rashdall (1936: 511ff.) compares the colleges of Paris with those of Oxford in some detail.

widely and stereotypically used for that purpose in England. It is known from its charter that Jocius's college was first established within one part of an existing hospice, where it remained for its first few decades before acquiring its own building, so there can be no question, in that instance, of a borrowed architectural form at the time of the initial foundation. Nevertheless, European contact with the Islamic Near East continued to be intense and prolonged for another century, so there was plenty of opportunity for the architectural design to have been copied as well.

At least some of the earliest European colleges thus followed the architectural design of the cloister, a rectangular building consisting of arcaded vaulted passages around an open courtyard. Like the Central Asian *vihāra* and *madrasa*, the early English college cloister had doors along the vaulted corridor opening onto rooms for the students and masters. Examples can still be seen at Oxford, such as the one in Magdalen College.[59]

The cloister plan is found not only in the old colleges but also in numerous monasteries, especially in France. It would seem, however, that the monastic cloisters are also no older than the twelfth century. Although the typical *madrasa* architectural plan was probably borrowed by the Western Europeans for the college and part of the monastery, there appears to be no study of the architectural origins of the medieval cloister.[60]

The reason for the persistence of the colleges at Oxford and Cambridge is that as both endowed and incorporated institutions governed by students and masters who belonged to them and were their lifelong beneficiaries, the latter had every reason to do their utmost to ensure the colleges' survival and the growth of their endowments. This was unlike the colleges at Paris, most of which eventually lost influence over time partly because the scholars who were supported by the endowments, whether students or masters, received simple payments administered by others, and partly because the *universitas* usurped the rights of the

[59] This is based on my personal visit in 2008. There are other examples at Oxford and elsewhere. See the next note.

[60] I say "apparently" and "seems" because I have not been able to find any scholarship specifically on the architecture of the early English colleges, or on the origins of medieval European cloisters in general. In addition to the well-known early college cloisters still in existence in Oxford and Cambridge, the Cloisters, a branch of the Metropolitan Museum of Art in New York City, includes several actual medieval French monastic cloisters that were dismantled and reassembled in New York in the mid-twentieth century; see the website at www.metmuseum.org/cloisters. For photographs of medieval French, Italian, and Spanish monastic cloisters in situ, see www.marcuslink.com/travel/cloisters/FR-stemilion.htm. It is doubtful that any are earlier than the thirteenth century, but this should be investigated by specialists in the history of art and architecture. The special issue of *Gesta* for 1973 features studies of much later cloisters in Europe.

colleges and in some cases simply took them over, as noted above. In Paris, the payments and affiliation ended as soon as the scholars left the college, so students did not develop the loyalty and affection for their college seen in England.[61] Nevertheless, many of the colleges survived down to the time of the French Revolution, when the entire University of Paris was closed by the revolutionary government.

The theology masters at the colleges of the Sorbonne and of Navarre lectured there—not in the public lecture halls—and their lectures were considered regular university instruction. By 1445, the University of Paris declared that "almost the whole University resides in the Colleges," where most teaching was by then carried out.[62] Later European universities mostly followed the developed Parisian college-university model, in which the university and the colleges had acquired similar functions and legal status, so that the terms 'university' and 'college' became essentially synonymous. Eventually, the university subsumed the college in most of Europe. As a result, scholars have had the false impression that there was a linear, rather miraculous development from the early European "university" in Bologna and Paris, which was a type of guild, to the typical later European university, which is rather like a college but oddly *sans* colleges. This makes Oxford, and the existence of the college in the High Middle Ages, enigmas. They are only enigmatic, however, if it is presumed that the university *as we know it* must be a purely European invention, just as the recursive argument method must be a purely European invention. This attitude is the reason why the early history of the college—which is almost completely foreign, and specifically Central Asian, in origin—has successfully eluded explanation, despite the admirable attempts of Makdisi to explain it.

Further support for this argument is given by Makdisi's discussion of Medieval Latin educational ranks or titles, several of which he shows are undoubtedly calque translations from Arabic.[63] Among the more striking of his examples is the *licentia docendi* 'license to teach', which first appears in Latin Europe in the late twelfth century, in a decree of Pope Alexander III (r. 1159–1181). It corresponds exactly to the Arabic *ijāza li-'l-tadrīs* 'license to teach', which had appeared in the Islamic world by about the tenth century.[64] In fact, the correspondence of these and

[61] Makdisi (1981: 236–237), Rashdall (1936: 511ff.).

[62] Rashdall (1936: 518ff.).

[63] Makdisi (1981: 270–280; 1990: 26–38).

[64] Makdisi (1981: 272).

other terms was already demonstrated beyond any reasonable doubt by nineteenth-century scholars, who argued that the Latin licentiate derives from the Arabic equivalent.[65] This strongly suggests that not only was the Latin college borrowed directly from the Islamic *madrasa*, which was in origin the Central Asian Buddhist *vihāra*, but to some extent the European higher educational system as a whole was derived from the medieval Islamic one. This should hardly be surprising, considering the massive cultural influence of the Islamic world on Latin Europe in this period in general.

[65] See Makdisi (1981: 274–275), who quotes some of this earlier scholarship, which he follows on this point. He notes that these scholars' work was all summarily dismissed by the medievalist Powicke (1879–1963), who simply pronounced his judgement that their work was "not convincing" (Makdisi 1981: 275). And there the matter has lain for another half century since Powicke's death. Makdisi politely suggests that the reason scholarly investigation of this question "made no progress beyond this point" may have been some shortcoming or other in the focus of Islamicist scholars, but the truth is that Powicke effectively silenced many good scholars for a century.

## — Chapter Four —
# Buddhist Central Asian Invention of the Method

There does not seem to be anything like a recursive argument in the Buddha's own teachings, as far as we can now tell. Because there are no written records of actual Buddhist texts for several centuries after the Buddha's death, little can be said for certain about him. Nevertheless, some Buddhist teachings, which are found in all Buddhist traditions—indicating that they may be inherited from early Buddhism—contain embedded sets in the form of linked lists, which are suggestive of recursion. It is also emphasized that the Buddha reached enlightenment specifically by analytical thinking. His feats of the mind are celebrated in later Buddhist literature, which treats them as what might be called "heroic thought." However, overt fully recursive argumentation seems otherwise to be absent in early Buddhist texts. It is only centuries later, in the Graeco-Bactrian and Graeco-Gandharan branches of the Sarvāstivāda school of Buddhism—that is, in Central Asia—that the recursive argument method developed out of their own special analytical approach.

The Buddhist fondness for explicitly numbered sets and lists is traditionally ascribed to Siddhārtha Gautama, or Śākyamuni Buddha, himself. In the Discourse at Vārāṇasī the Buddha is said to have given his first public presentation of the *Four Noble Truths*[1] and the *Eightfold Path*, both of which contain some of the basic insights he attained in his Enlightenment:[2]

[1] Also rendered as 'the four truths of the Noble ones'. Gethin (1998: 60) says, "The temptation to understand these four 'truths' as functioning as a kind of Buddhist creed should be resisted; they do not represent 'truth claims' that one must intellectually assent to on becoming a Buddhist. Part of the problem here is the word 'Truth'. The word *satya* (Pali *sacca*) can certainly mean truth, but it might equally be rendered as 'real' or 'actual thing'. That is, we are not dealing here with propositional truths with which we must either agree or disagree, but with four 'true things' or 'realities' whose nature, we are told, the Buddha finally understood on the night of his awakening." Some scholars specializing in early Buddhism doubt that there were four of them. Many texts refer to three or even fewer "truths"; see, e.g., Frauwallner (1957) and De Jong (1993).

[2] From Lamotte (1988: 26–27). I have added the material in square brackets, and the list of the five *skandhas*; the formatting is mine also.

### *The Four Noble Truths*

[I] This, O monks, is the noble truth of [the existence of] suffering:

[1] birth is suffering,
[2] old-age is suffering,
[3] disease is suffering,
[4] death is suffering,
[5] union with what one dislikes is suffering,
[6] separation from what one likes is suffering,
[7] not obtaining one's wish is suffering,
[8] the five kinds of objects of attachment[3] are suffering.

[These FIVE SKANDHAS or 'aggregates' are:

[1] corporeality (*rūpa*),
[2] feeling (*vedanā*),
[3] perception (*saṃjñā*),
[4] volition (*saṃskāra*), and
[5] consciousness (*vijñāna*).[4]]

[II] This, O monks, is the noble truth of the origin of suffering:

It is the thirst which leads from rebirth to rebirth, accompanied by pleasure and covetousness, which finds its pleasure here and there: the thirst for pleasure, the thirst for existence, the thirst for impermanence.

[III] This, O monks, is the noble truth of the cessation of suffering:

The extinction of that thirst by means of the complete annihilation of desire, by banishing desire, by renouncing it, by being delivered from it, by leaving it in place.[5]

[IV] This, O monks, is the noble truth of the path which leads to the cessation of suffering:

It is the NOBLE EIGHTFOLD PATH with eight branches called:[6]

[1] right faith,
[2] right will,
[3] right speech,

[3] Lamotte (1988: 26) adds here "(*upādānaskandha*)."
[4] Summarizing Lamotte (1988: 28), q.v. for the details.
[5] "Nirvāṇa is the subject of the third noble truth" (Lamotte 1988: 40).
[6] Lamotte (1988: 27) reads, "it is the noble path, with eight branches, which is called . . ."

[4] right action,
[5] right livelihood,
[6] right effort,
[7] right mindfulness and
[8] right concentration.

This predilection for listing might not be specifically Buddhist in origin, as it is found in Indian literature claimed to be pre-Buddhist, but it is remarkable that the basic teachings of Buddhism from a relatively early period on consist of linked, explicitly numbered lists. Although most of these teachings are found in every Buddhist tradition and are thus reconstructable to a relatively early period of organized Buddhism in India,[7] they are not reconstructable to the Buddha himself, and many scholars now believe the lists to be a later development.[8] In one text the Buddha is presented as saying "that he has always made known just two things, namely suffering and the cessation of suffering."[9] Based on what are thought to be the very earliest texts, it is necessary to add the teaching of the Middle Way leading to nirvāṇa (the third Noble Truth in the traditional list), but this still does not amount to a list in which some items consist of sublists, as in their traditional form. Moreover, it is uncertain when the use of linked lists or sets of lists became regularly marked by a numeral indicating the number of members in the list, for example, "the twelve bases of consciousness," "the five hindrances," and so on, though this too is widespread among Buddhist traditions.

Whether or not the Buddha used numbered lists, the practice of organizing his teachings in numbered lists certainly did come into being after his lifetime when the teachings had not yet been written down but still

[7] There is absolutely no evidence or indication that any Buddhist texts were actually written (or written down) until long after the Buddha's Nirvāṇa, but plenty of evidence that traditions about his teachings were regularly recited orally.

[8] According to Bareau and other scholars cited in De Jong (1993: 17–21), the numbered lists are undoubtedly later; cf. chapter 7. Similarly, Hirakawa (1990: 28) notes, "The Four Noble Truths . . . are designed to be used in instructing others and do not seem to represent the contents of the Buddha's enlightenment in its earliest form." He adds, "the twelve-link version of [the theory of] Dependent Origination [i.e., chain of causation] may be a systematized explanation based on Śākyamuni's meditations when he realized enlightenment." The systematization is unlikely to have gone back to the Buddha himself. As Gethin (1998: 64) notes, "If one were to define Buddhism . . . it would have to be as a practical method for dealing with the reality of suffering," or *duḥkha* 'uneasiness, difficulty, sorrow, pain', which refers to "an underlying sense of 'unsatisfactoriness' or 'unease' that must inevitably mar even our experience of happiness" (Gethin 1998: 61); it is the "burning" of *duḥkha* that is "blown out, extinguished" by nirvāṇa.

[9] Gethin (1998: 59). However, this could well be a later fabrication.

were passed on orally from teacher to pupil. It is thought that the practice of listing is helpful for oral teaching and memorization,[10] which was long the only method of transmission of knowledge known to have been used in India.[11] The numbering of the sets and lists is significant in that it may reflect the traditional strong interests of the Indians in numbers and mathematics in general. However, the use to which this practice was put in Buddhism seems on the whole not to have developed beyond the stage described here. It must also be stressed that this kind of listing is not recursive, and true recursive method arguments are not found in early Indian texts of any kind.[12]

Moreover, the enduring influence of Hellenism after Alexander's conquest and colonization of Bactria and Gandhāra, including what is now northwestern Pakistan, must be taken into account. In Bactria, the Greek script early became the normal script used to write the local Iranic language, Bactrian, and continued to be so used until the extinction of the language at the end of the first millennium AD, when Arabic and New Persian replaced Bactrian. Very recently, an inscription dedicating a stupa, dated 714, and fragments of Buddhist texts written in Bactrian, using the Bactrian form of Greek script, have been found.[13] The earliest Buddhist texts that have been found in the region are written in Kharoṣṭhi script, which probably derives from a form of imperial Persian Aramaic script; the language is Gāndhārī Prakrit, a Middle Indian dialect native to what is now southeastern Afghanistan and northwestern Pakistan.[14] The later Buddhist texts are in Sanskrit, written in Kushan Brahmi or other forms of Brahmi script.[15]

[10] Cf. Gethin (1998: 39).

[11] Writing, though introduced perhaps as early as the sixth century BC (with the Persian conquest), was little used until well into the common era.

[12] I have not found any examples earlier than the Central Asian scholastic method. See below in this chapter on putative earlier examples of the method in non-Buddhist texts.

[13] Van Bladel (2010: 54–55)

[14] For a long time, almost all of the early texts and inscriptions found in the region belonged to the Sarvāstivāda school, supporting literary and historical evidence. Recently, a large number of early texts written in Gāndhārī Prakrit were found. They belong mostly to the earlier, little-known Dharmaguptaka school, which was important in the transmission of Buddhism to China; the texts of the school otherwise exist almost exclusively in Chinese translation. See Salomon et al. (1999). Although Sanskrit did eventually displace Gāndhārī Prakrit for literary purposes, the earliest Sanskrit Buddhist philosophical manuscript so far discovered, known as the Spitzer Manuscript, which is mostly devoted to *abhidharma*, was found in Turfan (Turpan), in northeastern East Turkistan, and is provisionally dated to between AD 150 and 250 (Franco 2004: 32–33).

[15] Van Bladel (2010: 56).

## Scientific Thinking in Bactria and Gandhāra

The people in this same region must have known Greek medicine, because it has recently been shown to have been transmitted to Tibet during the Tibetan Empire period[16] (ca. AD 600–842), as the later Tibetan sources claim,[17] and it almost certainly came to Tibet from the Bactria-Gandhāra region, into which the Tibetans expanded for a time in the eighth and early ninth centuries.[18] It is also well known that there was considerable Greek influence on the artistic culture of Bactria and Gandhāra.[19]

However, the evidence would not seem to indicate any retained or transmitted knowledge of Aristotle or Greek in Central Asia after the conquest of the Greek kingdoms by the Kushans. Specifically, there is no known mention of Aristotle in Central Asian Buddhist scholastic texts. Moreover, a full scientific culture failed to develop in Central Asia or India before Islam despite knowledge of Indian mathematical science,[20] and despite the early presence and pervasive influence of both the college and the recursive argument method. Nevertheless, there is some evidence supporting the possibility of ancient Greek philosophical influence on Central Asian Buddhism.

The Sarvāstivāda school developed specifically in Bactria and Gandhāra, territories east of Persia that were at one time part of the Persian Empire conquered by Alexander the Great and settled with Greeks.[21] The region continued to be strongly Hellenized at least until the influx of a new Central Eurasian influence when the region was conquered two centuries later by a federation of three peoples: the Tokharians, *Aśvin, and Sakas, led by the Tokharians.[22] Out of their state the Kushan Empire developed. The Kushans evidently patronized all religions practiced in their empire, but they especially favored Buddhism.

The question of early Greek influence on Central Asian Buddhism remains little investigated, perhaps because it is complex and calls for a scholar able to work with a rare combination of difficult languages

[16] Yoeli-Tlalim (2010).

[17] Beckwith (1979). The one unquestioned Greek physician among those mentioned by name in the Tibetan sources is *Galenos* 'Galen' (Greek Γαληνός Galēnos).

[18] Beckwith (1993). Manuscripts of Buddhist medical works in Sanskrit have also been found in former Tokharistan close to Bāmiyān (Van Bladel 2010: 56).

[19] Some of this Greek influence eventually reached as far east as China; see Brooks (1999).

[20] This is assumed on the basis of fragments of various Indian scientific works found in the Kucha area (Sander 1999: 91).

[21] Indian influence on Greek thought (and vice versa) needs more study.

[22] Beckwith (2009: 84–85).

and subject matter,[23] but some preliminary observations might not be out of order. The unusual characteristics of the Sarvāstivāda school, in comparison with other early Buddhist schools, suggest that something different, something non-Indian, contributed to its development into its attested form. Among the remains of Greek texts discovered in the ruins of the Graeco-Bactrian city of Ay Khanum are part of a philosophical work of the Aristotelian school.[24] Although the discoveries by themselves are insufficient support for a theory of Greek influence on Sarvāstivāda Buddhism,[25] it is conceivable—in view of the long continuation of Greek medical traditions in the region—that certain Greek philosophical ideas or ways of thinking too may have survived there, even after knowledge of the literary tradition itself was lost.

One possible piece of evidence is textual. Barnes[26] and others translate Aristotle's various terms for arguments opposed to his own view as *objection*. However, this is not faithful to the Greek, in which the corresponding terms (if there are any in the text that do correspond) mean 'obstacle', 'difficulty'.[27] In this connection it may be noted that the Aristotelian term ἀπορία 'difficulty' corresponds to the frequent Chinese translation 難 'difficulty' of one of the terms in the *Mahāvibhāṣa* for what has traditionally been called an 'objection' (i.e., a 'subargument' in the terminology used in the present book). Moreover, because the latter text is modeled on the earlier Bactrian-Gandhāran *Vibhāṣa*, in which the earliest fully developed recursive argument occurs (see Example 4.1), it would seem possible that the similarities with Greek ideas—though not actual Greek disputational practice—might not be accidental.

Another possible bit of evidence has to do with the defining feature of the Sarvāstivāda school's analytical approach to Buddhism, with its emphasis on *abhidharma* (Buddhist scholasticism), especially its "realistic atomism": *sarvāstivāda* means, literally, 'the school of those who say everything really exists'. It is not known if 'everything' originally referred simply to the world in general, or already specifically to *dharmas*, atomistic 'constituents of phenomena'. The school's name is now interpreted to mean, 'those who say all [*dharmas*] really exist [in the past, present, and future]'; but taken in the simple sense, which could well be the original sense, it corresponds to Aristotle's 'realist' worldview. The development

[23] The texts of Sarvāstivāda Buddhism are mostly preserved only in Chinese translation.

[24] Bernard (1978: 456–460); Lerner (2003); Rapin (1992: 115–130).

[25] For some interesting historical material and arguments, see McEvilley (2002).

[26] E.g., Barnes (1984, 1: 422).

[27] In some cases so translated in Barnes (1984, 2: 1572 et seq.).

of a realist ontology involving atomistic constituents (which are, however, explicitly rejected by Aristotle) would thus be a later development. On the other hand, while this atomistic view is ubiquitous in early Indian thought, it is unknown in the earliest Greek thought. Democritus (ca. 460–ca. 370 BC)—who is said to have traveled to India—is the foremost early Greek proponent of an atomistic theory; his other views are also extremely similar to some typically Indian philosophical ideas.[28] Atomism seems to have developed very early in India, including in Buddhism, so that it is perhaps more likely that the Greeks took their atomistic ideas from the Indians than the other way around, yet the way in which Central Asian Buddhists developed these ideas and made them the defining features of their school might suggest that the general Greek cultural influence long outlasted the Greeks themselves in the region.

Nevertheless, all this does not alter the fact that the *recursive argument method* has no classical antecedents in the Graeco-Roman world, India, or China. It first appears in Buddhist Central Asia. The earliest text so far identified that uses a primitive version of the method, and indeed uses it throughout the text, is the Central Asian *Aṣṭagrantha*.[29] In this work, each topic argument is followed by a list of arguments about it—usually a rather long list—and then they are repeated and disputed, one by one, in order.[30] By contrast, the later *Jñānaprasthāna*,[31] the Kashmiris' reworking of the *Aṣṭagrantha*, completely omits the first list throughout.[32] That is, the *Jñānaprasthāna* does not use the recursive argument method at all. It strictly follows the two-part Question : Answer format.[33]

The *Vibhāṣa*, a scholastic work of the Bactrian-Gandhāran branch of the Sarvāstivāda school dated possibly to the first century AD,[34] during the Kushan Empire, contains the earliest known example of what eventually became the fully developed recursive argument method. The method

[28] Diogenes Laertius (1925, 2: 444–445, 452–455).

[29] *Taishō* 1543. The title has also been reconstructed as *Aṣṭaskandha*, and other proposals are possible, but reconstructions of a hypothetical Sanskrit title on the basis of the attested Chinese title are somewhat imaginary, since the work was undoubtedly composed in Gāndhārī Prakrit, not Sanskrit. I have followed tradition and the most usual "reconstruction."

[30] Cf. Cox (1998: 223).

[31] *Taishō* 1544.

[32] Cox (1998: 223). The text was converted into Sanskrit and revised in around the second century AD by the Kashmiris, who renamed their version the *Jñānaprasthāna* (Willemen 2006: 6).

[33] This point, along with others noted below in this chapter, suggests that the Central Asian recursive argument method was foreign to the Kashmiris, and that they did not really understand it.

[34] Willemen (2006) does not give an estimated date. There were once many *vibhāṣas*, q.v. Cox (1998: 229ff.).

apparently thus developed specifically within the Bactrian-Gandhāran branch of the Sarvāstivāda school, and was only later partially adopted by the Kashmiri Vaibhāṣika sect of Sarvāstivāda.[35]

Examples of the recursive argument do not occur in earlier Buddhist texts, earlier non-Buddhist Indian texts, or earlier texts connected to other branches of the Sarvāstivāda school. The traditional Indian version of the recursive argument method is thought by Indologists to be quite "early" and also "pre-Buddhist,"[36] but this judgement turns out to be based purely on a fabulous tradition, which in this case—as in most others like it—has no reliable support, as may be deduced from the comments of Potter:

> The date of Gautama . . . , to whom the *Nyāyasūtras* is attributed, is variously estimated from as early as the 6th century B.C. to the second century A.D., the reason for the discrepancy being apparently that these *sūtras*, which achieved their present form around the time of Nāgārjuna and very possibly were fashioned by a chief architect of that period, are attributed to a traditional personage who must have lived very long ago, since he is known to the authors of the Vedas and epics which date back many centuries before Christ.[37]

It is thus abundantly clear that the *Nyāyasūtras*, the putative "earliest" Indian texts to talk about such argumentation, cannot be any earlier than Nāgārjuna, the earliest Indian writer to use the method in a preserved work. Nāgārjuna is traditionally dated to around the second century AD.[38] The virtual identity of the later Indian method with the earlier

[35] See Willemen (2006: introduction), and Willemen, Dessein, and Cox (1998).

[36] In Buddhist studies, and Indology in general, whatever is claimed to be "earlier" often cannot really be dated, it often cannot possibly be as early as it is said to be, and it is also invariably not demonstrably pre-Buddhist in cases where that is claimed. There is no way to avoid relative dating—I am guilty as well—but I have attempted to exercise as much care and trepidation as seemed reasonable in each case.

[37] Potter (1977: 3–4). According to Potter (1977: 4), the earliest commentary on the work is said to have been written by Vātsyāyana, whom he dates to AD 450–500. However, as Michael L. Walter (personal communication, 2011) comments, "I can't think of an important religious or philosophical text in the world that had to wait nearly 1,000 years for a commentary; by then the work would simply have been forgotten." In short, one must rule out any significantly earlier date for Gautama than that of Vātsyāyana's commentary.

[38] The traditional date of the *Nyāyasūtras* is an excellent example of the Indian tradition of projecting more or less all of their important texts back into hoary antiquity. Nāgārjuna (whose actual dates are unknown) seems to be the first actual *Indian* writer to use a form of the recursive argument method, in his *Vigrahavyāvartanī* (q.v. Bhattacharya et al. 1978).

Central Asian method of the *Aṣṭagrantha* indicates that the former preserves the primitive Central Asian method, which was introduced to India during the Kushan Empire period along with the *vihāra* when that pro-Buddhist Central Asian dynasty ruled over much of northern India. In India, the method became fossilized at an early stage of development, as happened also with the *vihāra*. By contrast, in Central Asia, their home, both the college and the recursive argument method continued to develop. Although the recursive argument method is not found in any verifiably earlier Indian texts, whether Buddhist or non-Buddhist, nor is there any explicit description of it in such works, the primitive recursive argument of the Central Asian *Aṣṭagrantha* is used in later Indian philosophical texts, including works by Nāgārjuna, Kamalaśīla, and other Buddhists.[39] It must be stressed that the topical *content* per se of a recursive argument—for example, metaphysics, theology, biology, or whatever—is unconnected to its formal, overt structure, nor is there any connection between the structure of a recursive argument, on the one hand, and logical texts, debate texts, and elements of formal logic used in Indian disputations, on the other. The great importance of the method for science is its formal, overt structure, its most salient characteristic,[40] which impels the author to examine a question from all directions, including hypothetical ones, drawing attention to the unknown, and encouraging speculation on possible alternative solutions to problems.[41]

One of the most obvious of the many special features of the recursive argument method is its use of explicit numbering. Numbering occurs very early in Sarvāstivāda texts, and only increases, generally speaking, over the period of development of the recursive argument. Because of the fortuitous preservation of early examples, the changing practice of numbering over a period of several hundred years can be observed in the very same argument found in several Central Asian Buddhist texts that are based directly on each other. It would therefore seem appropriate to devote a brief discussion to it before turning to discussion of the Central Asian Buddhist recursive argument method.

The *Abhidharmahṛdaya*, a very early *abhidharma* text, does not use the recursive argument method at all. The *Aṣṭagrantha*, which uses a

[39] See chapter 7. These texts are crucially important for the history of the recursive argument method, as well as for the history of Buddhism, and should be examined carefully by Buddhologists.

[40] See chapter 2 for analysis of the structure of the recursive argument and the structure of texts written according to the rcursive argument method.

[41] Grant (2009); see chapter 2.

primitive form of the method, does not number the subarguments. The earliest *vibhāṣa* preserved (in Chinese translation),[42] which is also the earliest *abhidharma* text to number the subarguments, is the Central Asian work known as the *Vibhāṣa*,[43] in what is apparently the earliest example of the intra-Sarvāstivāda argument about "the differences among the constituents in the three times." It numbers the first list of subarguments (SUBARGUMENTS$_1$): "As for [number] one, . . . As for [number] two, . . . As for [number] three, . . . As for [number] four, . . . " For example, the first list's last subargument is overtly numbered, "as for [number] four," and its argument is given in capsule form: "the fourth [claims] a difference of otherness." The second list (SUBARGUMENTS$_2$) is not numbered. The arguments of the first list are referred to in the second list by repeating their capsule forms, *without* mentioning the number, for example, "The one who [claims] a difference of otherness. . . "

The same argument is found in the *Abhidharmavibhāṣa*,[44] the now fragmentary earlier Chinese translation of the Sanskritized and greatly expanded version of the Central Asian *Vibhāṣa* produced by the Vaibhāṣikas in Kashmir under the name *Abhidharmamahāvibhāṣa* or simply *Mahāvibhāṣa*.[45] It repeats the argument in the earlier *Vibhāṣa* practically verbatim, despite the fact that the translations are of different works and were done by different scholars at different times. This text again numbers the first list of subarguments, using cardinal numerals rather than ordinals: "View one [一說, literally, 'one view', or 'the view of one'] . . . , View two [二說, literally, 'two view', or 'the view of two'] . . . , View three [三說, literally, 'three view', or 'the view of three'] . . . , View four [四說, literally, 'four view', or 'the view of four'] . . . " However, the preceding sentence's mention of the school having four masters with different views makes it clear that they are referred to in this list, so the passage is to be understood as meaning "the first master's view," "the second master's view," and so on. As in the *Vibhāṣa*, the subarguments of the second list refer to the subarguments of the first list by way of a brief summary of the argument in each: "The one who claims a difference [based on] nature [i.e., the first] says . . . The one who claims a difference [based on]

[42] Willemen (2006: 5).

[43] Chinese text: *Taishō* 1547, 7: 466b.7–28. See below for an outline of its version of the argument. In these titles, *vibhāṣa* means something like 'commentary'.

[44] Chinese text: *Taishō* 1546, 40: 295c.7ff. See below for an outline of its version of the argument.

[45] The three *vibhāṣa* texts are distinguished conventionally by minor variations in their titles. The largest and latest, the translation by Hsüan Tsang, is conventionally distinguished from the others by the title *Mahāvibhāṣa*.

characteristic [i.e., the second] says . . . The one who claims a difference [based on] state [i.e., the third] says . . . The one who claims a difference [based on] difference [i.e., the fourth] says . . ." After the argument a question is added about the names of the Sarvāstivāda masters discussed. The author replies, transcribing their names and also numbering them: "The first is named Dharmatrā[ta], the second is named Ghoṣa[ka], the third is named Vasu[mitra], [and] the fourth is named Buddhadeva." Although the *Abhidharmavibhāṣa* is believed to be simply an earlier, now partly fragmentary, translation of the same text as the *Mahāvibhāṣa*, the differences between these two texts in this quotation are striking.

In addition, Dharmatrāta's *Saṃyuktābhidharmahṛdaya* is believed to be the text later reworked by Vasubandhu as the *Abhidharmakośabhāṣya*, but it does not number any of the subarguments in either list. By contrast, the version of the argument quoted in Vasubandhu's work does number the subarguments. Unlike the earlier *vibhāṣa* texts, however, his source (apparently a different version of the *Mahāvibhāṣa* not preserved elsewhere) does not number the first list:

> The Venerable Dharmatrāta's view is that there is a difference [based on] nature (*bhāva*). . . .
> The Venerable Ghoṣaka's view is that there is a difference [based on] characteristic (*lakṣana*). . . .
> The Venerable Vasumitra's view is that there is a difference based on state (*avasthā*). . . .
> The Venerable Buddhadeva's view is that there is a difference based on [relative] difference (*anyathā*). . . .[46]

Instead, the second list (the SUBARGUMENTS$_2$) is explicitly numbered: "The first . . . The second . . . The fourth . . ." (see EXAMPLE 4.1.c). It is striking, and significant, that Vasubandhu's source omits the third SUBARGUMENT$_2$, the view of Vasumitra, from the second list. Instead, it is referred to explicitly in the final AUTHOR'S VIEW ARGUMENT section: "Therefore, among these four, the third is the best." This gives us the reason for the omission of the third SUBARGUMENT$_2$ in reply to the third

[46] *Abhidharmakośabhāṣya*, Chinese: *Taishō* 1545, 77: 396a.13ff., Tibetan: Derge *ku*: 239v°.4–240r°.7, Sanskrit: Pradhan 1975: 296–297. For a complete translation of this section, see Pruden (1989: 806–810).

SUBARGUMENT$_1$, namely that the author agrees with that view. This is the earliest example so far identified of this distinctive practice, which is de rigueur in the Medieval Latin version of the recursive argument method.

Finally, the argument in Hsüan Tsang's fully preserved Chinese translation of the Kashmiri *Mahāvibhāṣa*[47] is notable in that it omits any numbering in either list of subarguments. However, the concluding AUTHOR'S VIEW ARGUMENT section refers implicitly to numbering in its comment, "Only the third [master's] establishing of the [three] times is good . . . ," and the reason it is considered to be the best is discussed briefly. This fact is of great importance for the history of the text.

As already noted, in Vasubandhu's version of the same argument, the second list of subarguments *is* overtly numbered, though the third subargument—with which the authors whom Vasubandhu quotes agree—is omitted from the list. Because Vasubandhu's version of this argument and the version of it given in the integral *Mahāvibhāṣa* both refer to Vasumitra's view by number ("the third") only in the concluding AUTHOR'S VIEW (which explains why Vasumitra's explanation is "the best"), it is obvious that the version of the *Mahāvibhāṣa* used by Vasubandhu is earlier than the version preserved in Hsüan Tsang's translation.[48] This chronological priority is confirmed by the confusion introduced into the later version in the second list of subarguments.

The change from the earlier system, which overtly numbers the first list of arguments, to the later system, where the second list of arguments is numbered, as in the *Mahāvibhāṣa* recension quoted by Vasubandhu, is also significant. The later system corresponds exactly to the system found in most thirteenth-century Latin writers, who, if they number the subarguments at all, only number those in the second list. In addition, the omission of the subargument with which the authors agree is striking. This is what is done in all of the thirteenth-century Latin recursive method arguments discussed in chapter 6.

The content of the above-quoted argument, in particular its succinct summary of the variant views by four revered masters, undoubtedly explains its popularity. It is about the defining doctrine of the historical

[47] Chinese text: *Taishō* 1545, 77: 396a.13ff.

[48] The *Abhidharmakośabhāṣya* is alone in being preserved in a Sanskrit manuscript as well as in translations into Chinese and Tibetan, so the text could be established very well in a critical edition, a translation based directly on the Sanskrit could be done, and the translations that have been done into various languages could be revised. Unfortunately, all of these tasks still remain to be done.

Sarvāstivāda school: the existence of *dharmas* 'constituents' in the past, present, and future.

## Examples of the Central Asian Buddhist Recursive Argument

As discussed above, the earliest preserved *Vibhāṣa* is a scholastic work that is believed to have been written in Gāndhārī Prakrit as a commentary on the *Aṣṭagrantha*, which had been written in approximately the first century BC in Gandhāra.[49] In about the second century AD, under the patronage of the late Kushans, the *Aṣṭagrantha* was reinterpreted and "translated" (or rather, mechanically converted) into Sanskrit[50] by scholars of the Vaibhāṣika subsect in Kashmir, who retitled it the *Jñānaprasthāna*. This is the text on which the same scholars' *Mahāvibhāṣa* is a commentary, modeled on the earlier *Vibhāṣa* but very many times larger and more complex. In effect, then, the Bactrian *Aṣṭagrantha* and *Vibhāṣa* are the models for the later *Jñānaprasthāna* and *Mahāvibhāṣa*. The Kashmiri Vaibhāṣika views expressed in the latter text are the butt of criticism by Vasubandhu (ca. fifth century AD), a Gandhāran who belonged to the Sautrāntika subsect of the Sarvāstivāda school.[51]

The first argument analyzed below, the same one discussed above with regard to the changes over time in the numbering systems, presents the different views of four great early Sarvāstivāda masters on the problem presented by the basic doctrine of their school. The argument is found in a number of versions.

The earliest, from the Bactrian-Gandhāran *Vibhāṣa*, is given in EXAMPLE 4.1a.

[49] Willemen (1998: xi); Willemen (2006: 6). Cf. Cox (1998: 229ff.) for a discussion of different scholars' views.

[50] It is now believed that early Buddhist works were first composed (though not actually written, at first) in one or another Prakrit (Middle Indian dialect). The Pali tradition is the best-known example. When Sanskrit became the prestige language of Buddhism in late antiquity, apparently during the ascendancy of the Kashmiri Vaibhāṣika subsect of Sarvāstivāda, texts previously written in Gāndhārī and other Prakrits were converted, rather mechanically, into Sanskrit. This was a major factor in the origin of what has been called "Buddhist hybrid Sanskrit," but perhaps more important, it provided an opportunity for new variations, or significant changes, to be introduced into old texts. This seems quite clearly to be what happened with the Sarvāstivāda texts converted by the Vaibhāṣikas into Sanskrit.

[51] After the demise of the Kashmiri Vaibhāṣikas, the Bactrian-Gandhāran branches of Sarvāstivāda claimed the position of orthodoxy and referred to themselves collectively as the Mūlasarvāstivāda (Willemen 1998: xi–xii).

An expanded version of the argument is found in the Kashmiri Vaibhāṣikas' great summa, the *Mahāvibhāṣa*, which is partly known to have existed in at least four recensions;[52] examples are given here from three of them:

1. The version of the argument in EXAMPLE 4.1b is found in the earliest of them, known as the *Abhidharmavibhāṣa*, which is partly preserved in Chinese translation.

2. The version in EXAMPLE 4.1c is from another early recension of the same work known from quotations preserved in Vasubandhu's *Abhidharmakośabhāṣya*.

3. The version given in EXAMPLE 4.1d is from the latest text of the *Mahāvibhāṣa*, preserved in Hsüan Tsang's complete Chinese translation.

Finally, the argument in EXAMPLE 4.1e is from Dharmatrāta's *Saṃyuktābhidharmahṛdaya*.

### *Example 4.1a*

***SITAPĀṆI**
***Vibhāṣa***
Differences among the Constituents in the Three Times[53]

[I. MAIN ARGUMENT (QUESTION or TOPIC)]

[As for the question of the differences among the constituents in the three times,] there are four Sarvāstivāda views on this.

[II. SUBARGUMENTS$_1$ about the MAIN ARGUMENT in I.]

[1] The first [claims] a difference of nature (事 *bhāva*).
[2] The second [claims] a difference of characteristic (相 *lakṣana*).
[3] The third [claims] a difference of state (時 *avasthā*).
[4] The fourth [claims] a difference of otherness (異 *anyathā*).

[III. SUBARGUMENTS$_2$ about the SUBARGUMENTS$_1$ in II.]

[1] The one who [claims] a difference of nature . . .
[2] The one who [claims] a difference of characteristic . . .

[52] The fourth recension of the *Mahāvibhāṣa* is known only from a few fragments of one Sanskrit leaf discovered near Kucha in East Turkistan and now preserved in the Pelliot collection in Paris (Cox 1998: 233–234).

[53] Chinese text: *Taishō* 1547, 7: 466b.7–28.

[3] The one who [claims] a difference of state . . . [In our] view [we] say that this is the least confused . . .

[4] The one who [claims] a difference of otherness . . . This is the most confused . . .

Note that the *Vibhāṣa* version of the argument in EXAMPLE 4.1a has no AUTHORS' VIEW section, perhaps because the author's view is stated in the third $\text{SUBARGUMENT}_2$, but this indicates that the *Vibhāṣa* belongs between the primitive and fully developed stages of the recursive argument method. Moreover, the text does not provide the names of the masters whose views are cited. The names are only cited later, in versions of the same argument found in the *Abhidharmavibhāṣa*, *Mahāvibhāṣa*, and *Abhidharmakośabhāṣya*, which are given below.

In 628 or 630,[54] the Chinese scholar monk Hsüan Tsang visited the great New Vihāra (Sanskrit *Nava Vihāra*,[55] Arabic *Nawbahār*) in the city of Balkh (ancient Bactra), in what is now northwestern Afghanistan. He stayed for over a month in that important Sarvāstivāda institution, which he says was built by "a former king of this country." More significantly, he notes, "Among the masters who compose *śāstras* north of the Great Snowy Mountains [the Hindu Kush], only this *saṅghārāma*'s continue the good work."[56] There he studied the great Vaibhāṣika summa, the *Mahāvibhāṣa*, with the master Prajñākara,[57] and acquired a copy of the work, which he translated when he returned to China. A massive book,[58] the *Mahāvibhāṣa* is a purely scholastic work in which Vaibhāṣika sectarian views on points of doctrine are presented and opposing positions are disputed, as shown in EXAMPLES 4.1b, 4.1c, 4.1d, 4.1e (four

[54] According to Ch'en (1992: 42–53) he visited in 628; the traditional date is 630.

[55] Hsüan Tsang (2000, 1: 117) uniquely calls it the *Nava Saṅghārāma*, which by his time was synonymous with *Nava Vihāra*.

[56] Hsüan Tsang (2000, 1: 117), who does not say anything nearly as detailed or complimentary about the large Buddhist *vihāras* he visited in the Tarim Basin region (East Turkistan), on which see Sander (1999) and Hartmann (1999). So far no cache of Buddhist manuscripts has been found in the vicinity of Balkh (though many secular manuscripts in Bactrian have been found near there), but many texts in Gāndhārī Prakrit have been found in the territory of the ancient kingdom of Gandhāra (now southeastern Afghanistan and northwestern Pakistan); see Salomon et al. (1999) and other works in the series published by the University of Washington Press.

[57] Li (1995: 48–49); the text also mentions two additional masters by name, and specifies several other Buddhist scholastic works in which Prajñākara was learned.

[58] It is the largest in the Chinese canon and takes up an entire thick volume of the *Taishō* edition.

textually different versions of the argument in 4.1a), and 4.2.[59] The rhetorical organization of the *Mahāvibhāṣa* and the form of its arguments are similar to those in other related scholastic texts, such as Vasubandhu's *Abhidharmakośabhāṣya*, a later work, as shown in EXAMPLES 4.3a and 4.3b.[60] The Sarvāstivāda recursive argument method has the same structure as the medieval Arabic and Latin methods.

### *Example 4.1b*

***Abhidharmavibhāṣa***

Differences among the Constituents in the Three Times[61]

[I. MAIN ARGUMENT (QUESTION or TOPIC)]

Among the Sarvāstivādins there are four *śāstra* masters who differ [in their views on the constituents in the three times].

[II. SUBARGUMENTS$_1$ about the MAIN ARGUMENT in I]

[1] The first view[62] is that the difference is [based on] nature.
[2] The second view is that the difference is [based on] characteristic.
[3] The third view is that the difference is [based on] state.
[4] The fourth view is that the difference is [based on] difference.

[III. SUBARGUMENTS$_2$ about the SUBARGUMENTS$_1$ in Part II]

[1] The one who claims a difference [based on] nature says . . .
[2] The one who claims a difference [based on] characteristic says . . .
[3] The one who claims a difference [based on] state says . . . [63]
[4] The one who claims a difference [based on] difference says . . .

[59] For a full translation and analysis of the part of the *Mahāvibhāṣa* in which this argument occurs, see Takeda and Cox (2010).

[60] Cf. the *Nyāyānusāra* (Cox 1995: 189–197), the Vaibhāṣika work written to refute the *Abhidharmakośabhāṣya.*

[61] Chinese text: *Taishō* 1546, 40: 295c.7ff.

[62] On the text's use of cardinal rather than ordinal numerals here, see the discussion of numbering above. My translation is based on the usual interpretation of 說 *shuō* in such contexts as meaning 'view, theory' rather than its modern meaning 'speak, say'.

[63] As in the *Vibhāṣa*, there is still no AUTHOR'S VIEW or CONCLUSION. The authors express their view in the third SUBARGUMENT$_2$, with the position of which they agreed.

After the argument a question is posed about the names of the masters who were discussed. The author then gives their names, which are transcribed phonetically (mostly in shortened form) in Chinese:

The first [master] is named Dharmatrā[ta], the second is named Ghoṣa[ka], the third is named Vasu[mitra], the fourth is named Buddhadeva.

### *Example 4.1c*

### *Mahāvibhāṣa*

(Quoted in Vasubandhu's *Abhidharmakośabhāṣya*)[64]
Differences among the Constituents in the Three Times[65]

[I. MAIN ARGUMENT (QUESTION OR TOPIC)]

How many differences are there among the [masters of] this school? Whose establishment of the [constituents of the three] times is the best and can be relied on? . . .

[II. SUBARGUMENTS$_1$ about the MAIN ARGUMENT in I]

[1] The Venerable Dharmatrāta's view is that there is a difference [based on] nature (*bhāva*).
[2] The Venerable Ghoṣaka's view is that there is a difference [based on] characteristic (*lakṣana*).
[3] The Venerable Vasumitra's view is that there is a difference based on state (*avasthā*).
[4] The Venerable Buddhadeva's view is that there is a difference based on [relative] difference (*anyathā*).

[III. SUBARGUMENTS$_2$ about the SUBARGUMENTS$_1$ in II]

[1] The first . . . [SUBARGUMENT$_2$] . . .
[2] The second . . . [SUBARGUMENT$_2$] . . .
[4] The fourth . . . (SUBARGUMENT$_2$] . . .

[IV. AUTHORS' VIEW ARGUMENT]

Therefore, among these four, the third [3] is the best. . . .

[64] For a complete translation of this passage, see Pruden (1989: 806–810). As previously noted, this text is evidently a quotation from the *Mahāvibhāṣa*, but from a recension different from Hsüan Tsang's, which does not number any of the arguments. Note that the argument, including the AUTHOR'S VIEW ARGUMENT here, is that of the original quoted text, not Vasubandhu.

[65] Chinese text: *Taishō* 1545, 77: 396a.13ff.; Tibetan text: Derge *ku*: 239v°.4 –240r°.7; Sanskrit text: Pradhan 1975: 296–297.

### *Example 4.1d*

### *Mahāvibhāṣa*

Differences among the Constituents in the Three Times[66]

[I. MAIN ARGUMENT (QUESTION or TOPIC)]

The Sarvāstivāda school has four great *śāstra* masters, each of whose establishment of the [constituents of the] three times is different.

[II. SUBARGUMENTS$_1$ about the MAIN ARGUMENT in I]

[1] The Venerable Dharmatrāta's view is that there is a difference [based on] nature (*bhāva*).
[2] The Venerable Ghoṣaka's view is that there is a difference [based on] characteristic (*lakṣana*).
[3] The Venerable Vasumitra's view is that there is a difference [based on] state (*avasthā*).
[4] The Venerable Buddhadeva's view is that there is a difference [based on relative] difference (*anyathā*).

[III. SUBARGUMENTS$_2$ about the SUBARGUMENTS$_1$ in II]

[1] As for the one who claims a difference [based on] nature . . .
[2] As for the one who claims a difference [based on] characteristic . . .
[3] As for the one who claims a difference [based on] state . . .
[4] As for the one who claims a difference [based on] difference . . .

[IV. AUTHORS' VIEW ARGUMENT]

Only the third's establishing of the [three] times is good, because . . .

The reference to Vasumitra's view as "the third" in the AUTHOR'S VIEW ARGUMENT section in EXAMPLE 4.1d suggests that an earlier recension did number the arguments in one of the two SUBARGUMENT sections. This particular text, unlike all the other variants, adds short additional arguments about the views of Ghoṣaka [2] and then Dharmatrāta [1], after the discussion of Vasumitra [3]. These arguments are out of sequential order and throw the entire SUBARGUMENT$_2$ section (III) into confusion. It may be a textual issue, but how such an error could have occurred is difficult to imagine according to any model of text transmission so far proposed for Buddhist texts, including oral recitation and simultaneous

[66] Chinese text: *Taishō* 1545, 77: 396a.13ff.

multiple transcription. One way or another, the confusion appears to be another example of the Kashmiris' failure to understand or appreciate the recursive argument method.[67]

Finally, the same argument is found in Dharmatrāta's *Saṃyuktābhidharmahṛdaya*, which is now widely believed to be the model for Vasubandhu's *Abhidharmakośabhāṣya*. It is placed here because—strictly as far as its argument structure is concerned—it seems likely to be later than all the others.

### *Example 4.1e*

**Dharmatrāta**
***Saṃyuktābhidharmahṛdaya***
The Different Interpretations of 'Sarvāstivāda'[68]

[I. Main Argument (Question or Topic)]

Question: How many kinds[69] of Sarvāstivāda are there?

[II. Subarguments$_1$ about the Main Argument in I]

[1] One kind [says there is] a difference of nature.
[2] There is another that says [there is] a difference of characteristic.
[3] Another says [there is] a difference of state.
[4] Another says [there is] a difference of difference.

[III. Author's View Argument]

These are the four kinds of Sarvāstivāda.

[IV. Subarguments$_2$ about the Subarguments$_1$ in II]

[1] As for the "one kind [that says there is] a difference of nature" . . .
[2] As for the [one that says there is] a difference of characteristic . . .
[3] As for the [one that says there is] a difference of state. . . . This [view] is thus the undisordered establishment of the [three] periods of time.[70]
[4] As for the [one that says there is] a difference of difference . . .

[67] However, this thesis should be tested by more extensive comparison of examples of the scholastic method from this text with examples from the other recensions.

[68] *Taishō* 1552, 11: 961c–962a (Dessein 1999, 3: 579–580). The title is mine. For a translation of the full argument, see Dessein (1999, 1: 749–751).

[69] I.e., 'interpretation', but the Chinese explicitly says 種 'kind, type'.

[70] As in the *Vibhāṣa* and other versions of this argument, the third view—Vasumitra's—is the approved one.

The recursive argument is also used elsewhere both in the *Mahāvibhāṣa*, as shown in Example 4.2, and in the *Abhidharmakośabhāṣya*, as shown below in Examples 4.3.a and 4.3b.

### *Example 4.2*

### *Mahāvibhāṣa*[71]

[I. Main Argument (Question or Topic)]

Do the past and future [times] have assembled [constituents], like [those in] walls and other things in the present time, or are they unassembled, and dispersed from one another? If [either] is posited, what difficulty[72] is there?

[II. Subarguments$_1$ (about the Main Argument in I)][73]

[1] If they were to have assembled [constituents], like walls and other things in the present time,

[1.1] It may be said[74] [that] . . .
[1.2] It may be said [that] . . .
[1.3] It may be said [that] . . .
[1.4] It may be said [that] . . .

[2] If they [the past and future constituents] are unassembled, and dispersed from one another,

[2.1] [Two arguments from scripture]

[2.1.1] It may be said . . . As the sūtra says [scripture quote] . . . Thus, [past] things like this . . .
[2.1.2] It may be said . . . As the sūtra says [scripture quote] . . . Thus, [future] things like this . . .

[2.2] It may be said . . .
[2.3] When future constituents came to assemble in the present . . .

[71] Translated into Chinese by Hsüan Tsang; text: *Taishō* 1546, 76: 395b.19–396a.1. My much abbreviated translation departs somewhat from the interpretation by Takeda and Cox (2010), q.v. for a full translation. I have regularly translated Chinese 法 *fǎ*, (S. *dharma*), which is often translated by Buddhologists as 'factor(s)', as 'constituent(s)'.

[72] Chinese 失 *shī* 'lose, loss'; it is used interchangeably with 過 *guò* 'to pass; transgression' and 難 *nán* 'difficulty, problem'. The text itself explicitly refers back to at least some arguments as 'difficulties'. I have translated all three terms as 'difficulty' or 'difficulties' throughout.

[73] The text here comments, "The two both entail difficulties" (過 *guò* 'to pass; transgression', here translated as 'difficulties').

[74] For "It may be said (that)" Chinese has simply 云 *yún* 'say, says, said, saying, (etc.) [that]'.

[III. SUBARGUMENTS$_2$ (about the SUBARGUMENTS$_1$ in II)]]

[1] Some have made this argument that both past and future [constituents] are assembled, like a wall and other things in the present.

[1.1] [As for the first], saying . . . We reply: . . .
[1.2] [As for the second], saying . . . We reply: . . .
[1.3] [As for the third], saying . . . We reply: . . .
[1.4] [As for the fourth], saying: . . . We reply: . . .

[2] The critic says that past and future [constituents] are not assembled like present things; rather, they are dispersed from each other.

[2.1] [As for the first], . . . We reply . . .

[2.1.1] [As for the first part, on the past,] . . . We reply . . .
[2.1.2] [As for the second part, on the future,] . . . We reply . . .

[2.2] [As for the second], which says . . . We reply . . .

[2.2.1] There are also proponents of another view . . .
[2.2.2] There are also proponents of another view . . .

[2.3] [As for the third] . . . We reply . . .

[IV. AUTHORS' VIEW ARGUMENT]

Assembled objects of experience depending upon really existing things are provisionally said to exist; sometimes [when assembled] they exist, and sometimes [when dispersed] they do not exist. This is thus the reason that the present [Vaibhāṣika] school asserts both existence and nonexistence.

The recursive argument method is also used by Vasubandhu himself in his *Abhidharmakośabhāṣya*. Vasubandhu's views in this work mainly support the earlier views of the Bactrians and Gandhārans against the later Kashmiri Vaibhāṣikas' views, as he found them in his copy of the *Mahāvibhāṣa*. Altogether, the *Abhidharmakośa* and the *Abhidharmakośabhāṣya* (which were circulated both separately and together) quickly became the single most important *abhidharma* text, and essentially replaced most earlier works.

Traditionally, the *Abhidharmakośa* has been explained as a highly abbreviated summary in verse (*kārikā*) form of the main doctrinal points written in extenso in the *Mahāvibhāṣa*, the great Vaibhāṣika summa. The traditional author of the *Abhidharmakośa* is Vasubandhu. The story goes that at first, the Kashmiri Vaibhāṣikas considered Vasubandhu's

*Abhidharmakośa* to be an excellent summary of their views. However, they were shortly thereafter chagrined to discover that immediately after finishing the *kārikās*, Vasubandhu did an about-face and wrote the *Abhidharmakośabhāṣya*, a detailed, highly critical refutation of many of the Vaibhāṣika school's tenets that contains frequent ad hominem comments about various Vaibhāṣika masters.

Although this story is still repeated as if it were history, it strains credulity. Vasubandhu is said to have done all this in accordance with a clandestine plan, or to have written his commentary on the *kārikās* after converting to the Yogācāra school, which is in turn said to mean that Vasubandhu converted to Mahāyāna. However, it has been shown that the *Abhidharmakośabhāṣya* "is ultimately based on the Bactrian *Abhidharmahṛdaya*" of Dharmaśreṣṭhin—the earliest and most important text of the Bactrian branch of the Sarvāstivāda school—via Dharmatrāta's *Saṃyuktābhidharmahṛdaya*.[75] Vasubandhu therefore did not innovate very much, but took advantage of the opportunity of writing his revision of the earlier works to include detailed refutation of Kashmiri Vaibhāṣika views found in his copy of the *Mahāvibhāṣa*. From the perspective of the Central Asians, the Vaibhāṣikas were the ones who had innovated, which is evidently why Vasubandhu, a member of the Sautrāntika subsect, considered them to be wrong and criticized them.[76] The Central Asians argued against the Kashmiris and eventually won by default, but they also continued to study the Kashmiris' *Mahāvibhāṣa*. The Sarvāstivāda tradition flourished in Bactria and vicinity after Vasubandhu up to and well after the Arab conquest.

Vasubandhu's own argument structure in the *Abhidharmakośabhāṣya* is slightly different from that found in the *vibhāṣa* texts, including most clearly the two latest versions of the *Mahāvibhāṣa*, namely the one he quotes and the one translated by Hsüan Tsang. Two examples of Vasubandhu's argument type are quoted below, one in EXAMPLE 4.3a (the same argument within which he quotes argument 4.1c above in full) and one in EXAMPLE 4.3b. The most important substantial difference between his argument structure and that of the earlier texts is that Vasubandhu apparently has just two viewpoints on the same Topic within one full recursive method argument; the SUBARGUMENTS$_1$ represent the opponent's view, while his own refutations of them are given in

[75] Willemen (2006: 20). Cf. Dessein (1999, 1: liv): "His *Abhidharmakośa* is based on the Abhidharma system as it had been worked out by Dharmaśreṣṭhin and revised and enlarged by Upaśānta and Dharmatrāta." Kritzer (2005) has shown that Vasubandhu's *Abhidharmakośabhāṣya* does include Yogācāra material. This does not, however, mean that Vasubandhu "converted" to Yogācāra, not to speak of Mahāyāna.

[76] Cf. the *Nyāyānusāra* (Cox 1995: 189–197).

the SUBARGUMENTS$_2$. He also does not number the arguments in either list. In these two respects he follows essentially the same *pūrvapakṣa* : *uttarapakṣa* argument structure that is followed in the Central Asian *Aṣṭagrantha* and in the Indian recursive argument method discussed in chapter 7. However, he does have a clear MAIN ARGUMENT (QUESTION or TOPIC) section and an AUTHOR'S VIEW ARGUMENT section.[77]

### *Example 4.3a*

**VASUBANDHU**
***Abhidharmakośabhāṣya***[78]

Existence of Constituents in the Three Time Periods

To what object one is bound by a past, present, or future *anuśaya*[79]

[I. MAIN ARGUMENT (QUESTION OR TOPIC)]

Whether constituents[80] exist in the three time periods.[81] Do the *kleśas*,[82] and past and future objects, really exist?

[II. SUBARGUMENTS$_1$]

[a. SUBARGUMENTS$_1$ Part 1: Based on scripture]

[1] The Vaibhāṣikas maintain that past and future constituents really exist . . . [scripture quoted]

[2] The Blessed One implicitly teaches the same doctrine when he says . . . [scripture quoted]

[b. SUBARGUMENTS$_1$ Part 2: Based on reason]

[77] These differences and similarities should be investigated by specialists in this literature.

[78] Chinese text: *Taishō* 1558, 20: 104a–106b; Tibetan text: Derge 140 (*ku*), 476(238 v°).2ff. The English text given here is from Pruden (1989: 806–820), with differences noted below.

[79] The term *anuśaya* is translated by Pruden (1989: 767) as 'latent defilement(s)'—a metalinguistic usage having virtually nothing to do with the English meaning of the words *latent* and *defilement*—but is usually left untranslated; cf. the following notes.

[80] I have silently converted Pruden's *dharma*(*s*) to 'constituent(s)' throughout. I have also corrected typographical errors and omitted the section and subsection numbers and page cross-references added in his translation.

[81] This sentence is my translation. The topic is given in the English translation (Pruden 1989: viii) as "Discussion: Do the *Dharmas* Exist in the Three Time Periods?" In fact, the argument occurs as a digression within the discussion of the *anuśayas* (or *kleśas*) 'latent defilements', in which the "three time periods" are raised in the text (Pruden 1989: 804): "We must examine to what object a person is bound by a past, present, or future *anuśaya*."

[82] The term *kleśa* is sometimes translated by Pruden as 'defilement'.

[3] A consciousness can arise given an object, but not if the object is not present. . . .

[4] If the past does not exist, how can good and bad action . . . Therefore, because of the proofs from scripture and reasoning quoted above, the Vaibhāṣikas affirm the existence of both the past and the future.[83]

[III. Author's View]

If the past and the future exist as things . . . [Two pages of argument (according to the Sautrāntika School) follow.] Consequently, . . .

[IV. Subarguments$_2$]

[a. Subarguments$_2$ Part 1: Based on scripture]

[1] The argument . . . is pure verbiage, for, . . .

[2] With regard to the argument that the Blessed One taught . . . we would also say . . .

[b. Subarguments$_2$ Part 2: Based on reason]

[3] As for the argument . . . Consequently the reason that the Sarvāstivādins gave . . . does not hold.

[4] The Sarvāstivādins also . . . But the Sautrāntikas do not admit . . .

[V. Author's View Argument]

Consequently, the *sarvāstivāda*, "the doctrine of the existence of all," of those Sarvāstivādins who affirm the real existence of the past and the future, is not good within Buddhism. It is not in this sense that one should understand *sarvāstivāda.* Good Sarvāstivāda consists in affirming the existence of "all" by understanding the word "all" as scripture understands it.[84] How do the Sūtras affirm that all exists? "When one

[83] Vasubandhu's "Commentary" here quotes verbatim the Sarvāstivādins' own arguments from the *Mahāvibhāṣa* about this essential point of their school. See Example 4.1.c for the quoted argument.

[84] Pruden's translation of de la Vallée Poussin's French translation goes back to the Chinese, so it is not surprising that the Sanskrit (which was found long after de la Vallée Poussin published his translation) often differs from it. In the Sanskrit this section reads, "So therefore, the Sarvāstivāda ['everything exists school'] is not correct (*sādhu*) regarding the teaching in saying 'the past and future exist substantially.' Yet [one who] says 'everything exists' as stated in the *sūtra* is correct." I am indebted to Richard Nance (personal communication, 2009) for help with this; all mistakes are of course my own. Vasubandhu belongs to the Sautrāntika "Scriptural" sect, so not surprisingly he often essentially argues that the Sarvāstivādin teachings are wrong on the basis of their misinterpretation of this or that sūtra quotation. This is a strategy used by theologians of medieval Islamic and Christian traditions as well.

says, 'all exists', Oh Brahmins, this refers to the twelve *āyatanas*: these are equivalent terms." . . . We have thus finished with the problem presented to us by the theory of the *anuśayas*.

### *Example 4.3b*

**Vasubandhu**
***Abhidharmakośabhāṣya***
Whether All Pleasurable Sensations Are Suffering[85]

[I. Main Argument (Topic)] The masters of certain schools deny that there is any real pleasure. They claim that it is only suffering, and say that this is known on the basis of scripture and reason.

[II. Subarguments$_1$]

[1] [Subargument$_1$ on the basis of scripture] . . .
[2] [Subargument$_1$ on the basis of scripture] . . .
[3] [Subargument$_1$ on the basis of scripture] . . .
[4] [Subargument$_1$ based on reason] What they say on the basis of reason [is] . . .
[5] [Subargument$_1$ based on reason] Again [又], . . . Therefore, all sensation is only suffering and it is proved that there is no real pleasure.

[III. Author's View Argument]

Opposing this, the *abhidharma* masters say that pleasure really does exist. . . .

[IV. Subarguments$_2$][86]

[1] [Subargument$_2$] As for [the argument that] the World-Honored One said there is no sensation that is without suffering . . .
[2] [Subargument$_2$] Again [又], as for [the argument that] the Sūtra says you should view all pleasurable sensations as suffering . . .
[3] [Subargument$_2$] Again [又], as for [the argument that] the Sūtra says . . . Thus it is proved that some real pleasure does exist. So also it is determined that their arguments for the nonexistence of real pleasure on the basis of scripture quotations do not work.

[85] *Taishō* 1558, 22: 114c.23–116a.9.

[86] The subarguments are not numbered in either list. Vasubandhu begins each Subargument$_2$ with a brief summary statement of the Subargument$_1$ it corresponds to, in all cases marked by the topic postposition 者 'as for', and then launches into his refutation. All of the Subarguments$_2$ are given in the correct order in the text.

[4] [SUBARGUMENT$_2$] Moreover [且], as for [the argument based on reason that] . . .

[5] [SUBARGUMENT$_2$] Again [又], as for [the argument based on reason that] . . .

[V. AUTHOR'S VIEW (Conclusion)]

Accordingly, it has been proved on the basis of reason that the sensation of pleasure really exists.

As mentioned above, Vasubandhu's own recursive argument is slightly different than the one in the version of the *Mahāvibhāṣa* that he quotes, and is reminiscent of the Indian type, though it is clearly a fully developed recursive argument method. It seems to be generally believed by most Indologists that the Indian form of the recursive argument is very ancient, in which case the Central Asian recursive argument method would presumably derive from it. However, as shown above, this chronological argument is unsupportable, since no actual or arguably earlier Indian texts that use such an argument structure are attested. The Central Asian *Aṣṭagrantha* is thus by far the earliest known text that uses a method like the Indian one. It is hardly conceivable that the Central Asians developed the recursive argument method, and later on Vasubandhu adopted the more primitive Indian argument type, but then updated it along the lines of the developed later Central Asian form. It would seem, rather, that Vasubandhu's method, while perhaps conservative (possibly due to influence from the simpler Indian method, as used by Nāgārjuna in his *Vigrahavyāvartanī*),[87] is nevertheless a version of the advanced Central Asian method, rather than a form of the Indian one derived from the primitive Central Asian type used in the *Aṣṭagrantha*.[88]

[87] See chapter 7.

[88] The Indian recursive argument method is in need of a major study.

## — Chapter Five —
# Islamization in Classical Arabic Central Asia

The Arab Empire founded by the prophet Muḥammad (d. AD 632) expanded rapidly, defeating the Byzantine Empire and capturing Syria (637) and Egypt (640). At the same time, the Arabs defeated the Sasanid Persian Empire (637) and raced across Persia into Central Asia, where they took Marw in 651, and Nishapur and Balkh in 652.[1] Soon after, they began the conquest of Transoxiana, establishing a lasting presence there within the next couple of decades, consolidating their control and expanding north to Khwārizm and northeast to Ferghana by 715. Almost simultaneously they moved through Sīstān (Sijistān, ancient Sakastāna, now southwestern Afghanistan), eventually capturing Sindh in 712–713.[2]

Within a very short time, then, early Arab Islamic culture came into direct, intimate contact with several major civilized areas, including the Graeco-Roman-influenced cultures of the Levant and North Africa, Persian culture, and the Buddhist cultures of Central Asia. From them the Muslims adopted various cultural elements. The question examined here is when, where, and how they acquired the recursive argument method and the college.

For much of the Umayyad Dynasty (661–750), the Arab Empire was centered on the dynasty's home territory, Syria, an ancient urbanized land that had been under Graeco-Roman rule for nearly a millennium and was thoroughly Hellenized. The dominant culture of the Near East in general, including Syria, had become Christian several centuries before Muḥammad. With the exception of medicine and other practical sciences, the intellectual achievements of classical antiquity had mostly been forgotten or were rejected as "pagan," though astrology-astronomy and alchemy continued to be studied in Syria and Egypt. The Arabs adopted much from their Christian and Jewish subjects, especially in the

[1] Bosworth (2008: 98), Beckwith (2009: 123).
[2] Wink (1990: 9).

field of medicine, which was thus fairly advanced in the central Islamic lands by the mid-eighth century.

Yet the intellectual center of Islam itself under the Umayyads was not in Damascus, the metropolis of Syria. It was not even in Syria at all. It was in Basra, a port town at the head of the Persian Gulf that very early became an Arab city and one of the great cities of the empire. The prosperity of Basra under the Umayyads was due partly to its political position as the administrative center and staging point for campaigns into the east, and partly to international trade, especially the trade with India in spices, aromatics, and silk.[3] Regular shipping connected the city to the ports of northwestern India, so that there was a constant presence of Indians in Basra, and Arabs in India.[4] Indian cultural influence, though evidently mostly oral, may have been significant in the formation of Arab Islamic intellectual culture.

Moreover, scholars flocked to Basra from elsewhere in the Arab Empire, enriching the city intellectually. Al-Mujāshi'ī (Al-Akhfash al-Ausaṭ, fl. ca. 795), the writer of the great early linguistic description of Arabic, *al-Kitāb* 'The Book',[5] was from Balkh, a city in Central Asia that at the time was still heavily Buddhist.[6] Most important, perhaps, Wāṣil ibn 'Aṭā' (d. 748) founded the Mu'tazila, 'the splitters' or 'Mu'tazilites',[7] the first philosophically grounded theological school in Islam, in Basra, where it was long influential. The Mu'tazilite school had at least one remarkable similarity in doctrine with the Vaibhāṣikas. The Mu'tazilites promoted a theory of particles ('atoms' or 'corpuscles'), ephemeral 'constituents' of the phenomenal world (including emotions and other nonphysical things),[8] and

[3] On the lucrative trade in musk and other aromatics, see King (2007, 2010).

[4] Although the heyday of Basra was the eighth and ninth centuries, it long continued to be the main Persian Gulf port for this trade, famously romanticized in the classic tales of Sindbad in the *Thousand and One Nights*.

[5] Al-Mujāshi'ī was a student of the grammarian Sībawayh, who is usually said to have written it (Sezgin 1984: 43–54, 68). It is believed to be based on notes taken by Al-Mujāshi'ī, though it presumably does largely represent Sībawayh's ideas, as tradition claims, probably from oral teaching.

[6] It has been suggested that some aspects of the book show Indian grammatical ideas. See Beckwith (2009: 413–414); cf. Danecki, cited in Van Ess (2002: 24 n. 29).

[7] The original meaning of *Mu'tazila* is unknown; the translation 'those who separate themselves, who stand aside' (Gimaret 1993) seems to be based on a folk etymology. The Tibetan translation of the name Vaibhāṣika is *byebragtu smraba* 'the splitter school'; the usual (perhaps original) explanation is 'those of the *Vibhāṣa* (commentarial texts)', or 'the commentators'. An explanation based on the "splitting" up of the phenomenal world into atomistic "constituents" would make sense for both traditions.

[8] They endure only for one moment. According to what became the orthodox Islamic view, they are constantly being created by God, and if He were to stop creating them, everything would cease to exist except Him. Their Buddhist equivalent, the *dharmas*, also exist only for the period of one

their 'accidents' or secondary features. The theory eventually came to be accepted as an orthodox view in Islam,[9] though it has no clear basis in scripture. It is also specifically non-Greek. The theory is widely considered to have Indian origins, and has been compared to the ideas about the same thing in the Buddhist Sarvāstivāda school and the Nyāya and Vaiśeṣika schools of Indian Brahmanism.[10] Another possible means of transmission of such ideas was of course via Bactria itself. Jahm ibn Ṣafwān of Tirmidh (d. 746), not far to the east of Balkh in Tokharistan, had debates specifically with Buddhists, on account of which he was forced to change his arguments on epistemology at least.[11] In short, although there are differences between the usual Islamic and Indo-Buddhist views, the fact remains that a theory of unstable "momentary" atomistic constituents with discrete attributes or "accidentals" was a fundamental view in both, and it is not original in Islam, the Near East, or the Graeco-Roman world.[12]

---

moment and are constantly coming into existence and going out of existence. See Von Rospatt (1998). The divine element is of course missing in Buddhism, arguing for a Brahmanical source, but see the following notes.

[9] It is sometimes called 'occasionalism'; see Fakhry (1958, 1983). The theory of atoms and accidentals eventually became orthodoxy because al-Ash'arī (873/874–935/936), founder of one of the most important of the orthodox schools of Islamic theology, was originally a Mu'tazilite. Despite his opposition to some Mu'tazilite positions, he retained most of their views largely unchanged when he founded his own school.

[10] See Von Rospatt (1998) for a careful, insightful comparison of one of the key features of this doctrine, the momentary existence of the constituents. Pines (1997) is the classic work on the subject. Unfortunately, European scholarship in 1936 (when Pines's book was first published in German) was not yet well enough informed about Indian philosophy, especially the main Central Asian Buddhist school, the Sarvāstivāda, to allow Pines to draw firm conclusions about precisely which school was the probable donor. This problem is to a large extent still with us; only in the past two decades has fundamental work been done on these schools. It was formerly popular to compare the Islamic system to the Nyāya and Vaiśeṣika schools of Brahmanism, which were much better known than the Vaibhāṣika and Sautrāntika subsects of Sarvāstivāda. However, for the early Islamic period a connection with Brahmanism is less likely than one with Sarvāstivāda Buddhism, which was still practiced and taught over a large swath of territory in the Arab Empire itself. With regard to the idea of divine creation of the constituents, Brahmanism might make better sense, in which case transmission via the commerce between Basra and the northwest Indian ports would be the logical answer. But a divine source is not central to the theory, and would in any case automatically be supplied by Islamic thought, according to which everything has a divine source, so divinity is nondistinctive and thus irrelevant here.

[11] Van Bladel (2010: 88).

[12] The article by Van Ess (2002) about research on the topic since the work of Pines discusses the Indian connection famously proposed by Pines, but oddly remarks, "Until recently, no Islamicist and no Indologist has dared approach the problem again" (van Ess 2002: 25). He argues essentially that it was an internal Arab Islamic development, though he suggests a Persian connection (which is unlikely in the extreme), and Greek influence (which he himself disproves for the early period).

Although the Muʿtazilites are also the first Muslims known to have used dialectical argumentation against their opponents,[13] so far none of the attested early examples of their argument structure—at least, in Muʿtazilite texts earlier than Ibn Sīnā (Avicenna)—are similar to the recursive argument method. They are thus not the initiators of the Arabic form of the method, and Basra was not where the connection and transmission took place.

The third point of significant cultural impact on early Islamic civilization was Central Asia. The long modern scholarly tradition of treating Central Asia as an eastern frontier zone of "Persian" (or, nowadays, "Iranian," usually in the sense of "Persian") culture, and Central Asians as, therefore, "Persians," is responsible for much misunderstanding of the region and its singular importance in the early history of Islamic culture.[14] Many key features of Classical Arabic civilization derive from Central Asian models. For example, the '*ghulām*' system, a special type of guard corps that became one of the distinctive features of Islamic governments before modern times, was adapted directly from Transoxanian Central Asia by the very first Arabs to go there in the seventh century.[15] The *madrasa*, or college, has been firmly identified by archaeologists, on the basis of actual excavated examples, as an Islamicized version of the Central Asian Buddhist *vihāra*. As discussed above, the distinctive "Islamic" version of atomism was probably adopted from the Central Asian Sarvāstivāda school of Buddhism, the defining metaphysical feature of which is precisely such an atomistic theory. Finally, it cannot be stressed enough that most of the great scholars and scientists of the classical age of Islamic civilization were not Arabs or Syrians or Persians; they were, overwhelmingly, Central Asians.[16]

The Nawbahār of Balkh long continued to function under Arab Islamic rule. It was damaged during the eighth-century wars in the

---

He cites Von Rospatt's (1998) article in a footnote, without comment; cf. similarly Dhanani (1994). The issue has yet to be discussed by anyone expert in both early Islamic philosophy and near-contemporaneous Buddhist philosophy, especially Sarvāstivāda. Cf. the preceding notes.

[13] See Fakhry (1983), still perhaps the best general treatment of early Islamic thought.

[14] This is especially true of work done within the field of "Iranian studies" down to the present day, but also of anything to do with the vast territory in which one or another "Iranian" language was spoken. Examples are legion. I have recently begun using 'Iranic' to refer to the language family, its members, and related cultures, for the sake of clarity.

[15] Beckwith (1984a); this is not made clear by de la Vaissière (2007). Moreover, the latter's idea that the Central Eurasian comitatus was a specifically Sogdian system, and all examples of it are evidence of Sogdians, is solidly disproved by the data, q.v. Beckwith (1984a; 2009).

[16] For a recent survey, see Starr (2009); cf. Beckwith (2009).

region, but it was repaired (in ca. 725–726)[17] and is described in some detail in the fragments of an account of the Barmakids by Abū Ḥafṣ al-Kirmānī (fl. ca. 800),[18] and in the account preserved in the anonymous *Ḥudūd al-'ālam* (dated AD 982).[19] Exactly when it ceased to function as a Buddhist *vihāra* is unknown. Very many Buddhist *vihāras* continued to operate in Khuttal as well, at least until 726, when the Buddhist monk Hui Ch'ao passed through the region.[20] Research has shown that the Buddhist center in Bamiyan, in the Hindu Kush south of Balkh, continued to prosper for some three centuries after the coming of Islam, well into the tenth century,[21] when the Ghaznavids exerted powerful pressure on non-Muslims throughout their empire to convert.[22]

The discontent that had long been building up in the Arab Empire finally burst into an open rebellion known as the Abbasid Revolution. It was led by Central Asian merchants, both Arab and non-Arab, who were based in Marw, the colonial capital of Arab Central Asia. The rebel army, the Khurāsāniyya or 'Central Asians', overthrew the Umayyad Dynasty and established a new dynasty, the Abbasid, in 750. One of the leaders of the revolution, Khālid ibn Barmak (d. AD 781–782), was the son of the Nawbahār's last known Buddhist rector, the Barmak, who had been educated in Kashmir.[23] Khālid was de facto vizier under the first Abbasid caliph, al-Saffāḥ (r. AD 750–754). He also advised the second Abbasid caliph, Abû Ja'far al-Manṣûr (r. AD 754–775) on the building of a new permanent capital, the City of Peace. A new location and new construction were needed because the Abbasids had not disbanded the army, which followed the caliph from temporary capital to temporary capital. Finally, a location was chosen near the Christian town of Baghdad on the Tigris River. Under Khālid's influence, the model chosen was the Central Asian circular palace-city plan,[24]

[17] Van Bladel (2010: 73).

[18] Text and translation in Van Bladel (2010: 62–66).

[19] For a translation, see the end of the main text in Beckwith (1984b).

[20] Fuchs (1938), Litvinskij (1985). For a valuable discussion of the early *ribāṭ* in Central Asia and its connection with earlier Buddhist institutions, see de la Vaissière (2008). Note, however, that the long-postulated connection between the *vihāra* and *madrasa* has been confirmed by the excavation of Adzhina Tepa, q.v. Litvinskij and Zeimal' (1971); cf. chapter 3. As de la Vaissière himself remarks, the *ribāṭ* does not have a Central Asian origin, so there is no direct connection between it and the *madrasa*, an originally Central Asian institution.

[21] Klimburg-Salter (2008).

[22] An aspect of the Islamic-Buddhist interface in the neighboring Karakorum Mountain region, interfaith marriage for political reasons in the postmedieval period, has been explored in an interesting study by Halkias (2010). It is probable that a similar situation obtained earlier in Central Asia; research on it is needed.

[23] Van Bladel (2010: 66, 69, 71–72).

[24] Beckwith (1984b; 2009: 147, 394 n. 28).

which had been used both for the former Parthian and Sasanid capital of Ctesiphon, about thirty kilometers southeast of Baghdad, and for the Nawbahār, a converted Parthian or Sasanid palace,[25] Khālid's ancestral home. He must have been familiar with both sites.[26] Construction began in 762 and al-Manṣûr moved to the palace city with his Central Asian army in the following year, when it was nearly completed.[27]

The new capital at Baghdad, as the capital of an enormous empire stretching from northwestern Africa to northwestern India, including nearly all of western and southern Central Asia, attracted ambitious men of all kinds. The city soon became the intellectual heart of the empire. Many scholars moved from Basra to Baghdad.

The Abbasid Caliphate as a whole was dominated for much of the first half century of the Abbasid Dynasty by Central Asians living in the capital, Baghdad. The most famous of them were the Barmakids, among whom the most important was Khālid's son, Yaḥyā ibn Khālid ibn Barmak (r. 786–803; d. 805), the famous vizier of Hārūn al-Rashīd. Yaḥyā fostered the introduction of Indian learning into Islam in the second half of the eighth century by bringing Sanskrit scientific books from India, along with Indian scholars to help translate them into Arabic. Under his direct, personal patronage, major works of Indian science were thus translated into Arabic in Baghdad before the large-scale project to translate the great works of Greek science began in the early ninth century.[28] The works translated from Sanskrit included:

i. The *Siddhānta* (AD 628) of Brahmagupta, the most advanced work of Indian astronomy in the eighth century, or possibly a larger work that included it. It was translated into Arabic by al-Fazārī in 770 or 772, during the reign of the caliph al-Manṣūr (r. 754–775), and became known as the

[25] Beckwith (1984b) treats the architectural plan of the Nawbahār and its description by Hsüan Tsang and in Islamic sources; how Khālid designed the new caliphal capital at Baghdad, the City of Peace, based on the Parthian-Sasanid circular royal palace plan; comparison with the *vihāra* of Samye in Tibet, which was built at about the same time, not two decades later as stated in my article, which follows the traditional chronology (on the date of Samye, see Walter and Beckwith 2010); and the shared elements in these circular plans, which are ultimately Central Eurasian in origin (*pace* Van Bladel 2010: 87). My article (1984b) must also be corrected to note that the Parthian city of Ctesiphon was originally circular in plan (cf. Van Bladel 2010: 87 n. 200).

[26] Khālid, as head of the Barmakid family of hereditary rectors, was personally in charge of the Nawbahār and its extensive endowed (*waqf*) property, on which see Van Bladel (2010: 64–67).

[27] Duri (1960). For a translation of the detailed description in Khaṭīb al-Baghdādī's *Ta'rīkh Baghdād*, see Lassner (1970); cf. the comments and additional information from other sources in Beckwith (1984b).

[28] A few texts were translated from Greek in the eighth century, but most claims about pre-ninth-century translations from Greek (or even Syriac) are very poorly supported and most should be rejected; see Sezgin (1978: 13ff.).

*Sindhind.*[29] It was immediately put to use by astronomers, though it was superseded soon afterward by the Hellenistic astronomy of Ptolemy, the translation of which was patronized personally by Yaḥyā.[30] The *Sindhind* is most important for its introduction of Indian mathematics, including some elements used by al-Khwārizmī (fl. AD 807–847), along with Greek mathematics, in his famous *Algebra*, which is his own synthesis of these elements in an already strongly Hellenized form.[31] His sine table also derives from the trigonometrical chapter of the *Sindhind.*[32] The *Sindhind* itself was reworked several times, most importantly by al-Khwārizmī, who based his *Zīj al-Sindhind* on it[33] but "Ptolemaicized" the Indian system.[34]

ii. The source or sources for Al-Khwārizmī's book on Indian numerals (called Arabic numerals in Europe) and mathematics using them, which is now known as the *Liber Algorithmi*,[35] must have been translated into Arabic at about this time.[36] Among other things, it includes explanation of the use of the zero, rules for basic decimal place system mathematics using Indian numerals (our "Arabic" numerals), and how to perform operations with fractions, including sexagesimal fractions, which were used in astronomical calculation.[37]

iii. The *Caraka-saṃhitā* and *Suśruta-saṃhitā*, the two most important works of Indian medicine, the *Aṣṭāṅgahṛdaya-saṃhitā* of Vāgbhaṭa, the *Siddhasāra* of Ravigupta, and several other Indian medical works were also translated into Arabic in the late eighth century.[38]

iv. Indian works of alchemy (or works connected to that relative of chemistry) would seem likely to have been translated at this time as well.[39] Though most of the works attributed to the greatest alchemist of all time

[29] Sezgin (1978: 116; 1974: 199); Van Bladel (2010: 83–84).

[30] Sezgin (1978: 84–85); cf. Van Bladel (2010: 85).

[31] Sezgin (1974: 228ff.).

[32] Sezgin (1974: 239).

[33] Vernet (1997).

[34] Van Bladel (2010: 84).

[35] Or *Algoarismi*, *Algorismi*, and other variants. The book title is the source of the words *algorism* and *algorithm*.

[36] On the problem of the sources for this work, see Sezgin (1974: 238ff.).

[37] Vernet (1997: 1070–1071).

[38] Sezgin (1970: 197–198, 238), Van Bladel (2010: 76). The *Suśruta-saṃhitā* was translated by the Indian physician Manka, who is also well known from historical accounts of the time (Sezgin 1970: 200–201).

[39] The earliest alchemist of the Arabic tradition, the Umayyad prince Khālid ibn Yazīd ibn Mu'āwiya (fl. early eighth century), who would have belonged to the Near Eastern tradition (Sezgin 1971: 120ff.), has been shown by Ullmann to be purely legendary (Gutas 1998: 24).

by reputation, Jābir ibn Ḥayyān (known to Latin Europe as Geber)[40] are evidently written by later scholars, one of the few clear historical facts about him is that he was closely connected to the Barmakids (one of whom was his student), such that he too suffered when they fell from power.[41] This lends support to the tradition that he was born in Ṭūs, in Central Asia, in the early eighth century;[42] in any case he seems to have died there in 815.[43] If he was raised in Ṭūs, it is practically certain that he would have received a Central Asian education, which at that time and place must have been either still largely Buddhist or heavily influenced by Buddhism, the centuries-old high culture in that region. Nevertheless, though the Jabirian corpus does include material of Indian origin, the works attributed to him for the most part grow out of the late Graeco-Roman tradition.[44]

During the eighth century, the Arabs' need for advanced medicine was largely satisfied by Syrian Christian physicians following the Greek medical tradition. According to scholarly tradition, they were based in a Sasanid teaching hospital and school then still operating in Jundaysābūr (located in what is now Khuzestan, in southwest Iran). However, there is now considerable doubt about that tradition.[45] Arabic sources actually claim that the medical tradition there went back to "a physician who had come from India."[46] When on Hārūn al-Rashīd's death the empire was divided into two parts according to his will, the West, with Baghdad and the position of caliph, was accorded to al-Amīn, while the East, with the provincial capital of Marw, was accorded to his brother al-Ma'mūn. This did not satisfy al-Amīn, who attacked his brother. The resulting civil war was won by al-Ma'mūn (r. 808/813–833) with the help of a Central Asian army. In the Central Asian city of Marw, his capital for ten years, al-Ma'mūn gathered a

[40] See Sezgin (1971: 132ff.) for a careful examination of his life and works.

[41] Sezgin (1971: 135, 221).

[42] The alternative tradition that he was a native of Ḥarrān in northern Syria (now southeastern Anatolian Turkey) is doubtful for various reasons, including the fact that *Ḥayyān* differs from *Ḥarrān* by only a single letter in Arabic.

[43] Sezgin (1971: 134). Tradition says he was born either in Kufa (in Southern Iraq) or in Ṭūs.

[44] Sezgin (1971: 174). It is accepted that Indian metal-based alchemy, and perhaps Indian alchemy in general, is fairly late and Western in origin.

[45] The accounts of the importance of Jundaysābūr (Gondeshapur) seem to be more or less completely legendary, other than as the location of a hospital and home of some influential early Christian physicians; see Lindberg (2007: 164–165). However, the attribution of importance to Persian literary culture in the development of Classical Arabic intellectual civilization, argued at length by Lindberg—buttressed by misattribution of Persian nationality to the Barmakids, among others—is incorrect.

[46] Sayili (1991: 1120), who notes that the Nestorian transformation of the town "may not have effectively taken place before the reign of Khusraw I Anushirawan (531–579 A.D.)."

brilliant intellectual court, which he brought with him to Baghdad, along with his army, when he finally returned there in 819.[47]

Al-Ma'mūn patronized learning, especially the translation into Arabic of Greek philosophical and scientific books, indirectly from Syriac and directly from Greek. The late eighth-century Central Asian and Indian cultural influence was thus followed immediately by an influx of Greek learning, which had begun already at the very end of the eighth century but became a virtual flood under al-Ma'mūn and his successors. This project made the majority of Greek scientific literature available in Arabic by the early tenth century. It completely reshaped Classical Arabic intellectual civilization. The translations from Greek (either directly, or indirectly via Syriac[48]) included:

i. The works of Aristotle
ii. The works of Plato
iii. The works of Hippocrates and Galen
iv. The works of Euclid
v. The works of Ptolemy
vi. The works of numerous other Graeco-Roman scientists

The greatest scientist of al-Ma'mūn's court was al-Khwārizmī (Algorithmus or Algorismus, fl. 807–847), discussed above. His works on mathematics, including Indian numerals and computation, were perhaps the most influential works in the new tradition of mathematics of early Classical Arabic civilization.

Although the early Central Asian and Indian influence on the formative period of Islamic civilization appears to have been significant, the new translations from Greek introduced the greatest intellectual achievements of classical Western civilization, including most of the works of Aristotle. The Greek works overwhelmed the earlier tradition, which was largely forgotten, but in several cases it seems that the Central Asian and Indian traditions remained part of the covert foundations of fields

[47] Beckwith (2009: 152).

[48] Claims that many Indian and Greek scientific works were translated into Arabic via Pahlavi (Middle Persian) intermediaries are unsupported by solid historical sources and must be rejected; the few clear exceptions are works of belles-lettres such as Ibn al-Muqaffa's Arabic translation, ca. 750, of "the Indian fable book and mirror for princes, *Kalīla wa Dimna*" (Gutas 2010: 19). Cf. the doubts of Van Bladel (2010), and his suggestion (supported by some historical sources) that if any Iranic intermediaries were involved they would have been Bactrians. Similarly, claims for unusually early dates of translation or composition of important philosophical or scientific works, and unlikely attributions to famous authors, need serious reexamination. For a fairly critical treatment, see Sezgin's *Geschichte des Arabischen Schrifttums*.

of knowledge that came to be dominated overtly by Greek thought. This was perhaps due in part to the fact that their origins had been forgotten, allowing their approaches, points of view, and even specific ideas, to be naturalized as integral parts of the new Classical Arabic civilization. Conversely, Greek science, coming as it did later, and obtained as it was in some cases directly from the Christian Byzantine Greeks—who were the enemies of the Muslims—was known to be "foreign," and therefore suspect and liable to be suppressed.

The translation of Aristotle into Arabic introduced the *idea* of science, which did not exist in Arabic before his works were translated, and of course it also introduced Aristotle's actual thought on specific topics, especially metaphysics and the natural sciences. The early, heady days of excitement over the discovery of Aristotle (however much neo-Platonized) continued for quite a while. The melding of Arab, Central Asian, Indian, and Greek thought brought about the brilliant flowering of Classical Arabic civilization.[49]

The early works of Classical Arabic science,[50] including those by al-Khwārizmī, were dependent on both Indic and Greek traditions. This continued under al-Ma'mūn's immediate successors, though the Indian tradition quickly receded.

1. The Christian physician (and later Muslim convert) 'Alī b. Sahl Rabban al-Ṭabarī (fl. ca. 820–860), a native of Marw in Central Asia, wrote an encyclopedic work on medicine, the *Firdaus al-ḥikma* 'The Paradise of Wisdom', which is largely Greek in orientation but includes a chapter covering Indian medicine. He was the teacher of the great physician al-Rāzī (Rhazes, q.v. below), probably directly, as stated by tradition, or perhaps indirectly via his writings.[51]

[49] Persian culture seems to have had virtually no influence at all on the development of Classical Arabic civilization outside the realms of politics and belles-lettres, despite the claims sometimes made about the supposed strong formative influence of Persian intellectual culture, including even Zoroastrianism, on early Islamic civilization. The references to putative Sasanid scientific works in Pahlavi (Middle Persian), whether native or translated, are mostly fantasies based on the most unreliable accounts imaginable. Attempts by some to resurrect a glorious Sasanid scientific past are supported by references to patently ahistorical works filled with legends and marvels and, not least, references to much later times. Cf. the preceding note.

[50] I omit those ascribed to the Umayyad period. As far as they are historical, they appear to have belonged to the Syriac tradition. The early alchemist and physician Khālid ibn Yazīd ibn Mu'āwiya is purely legendary; see this chapter, note 39.

[51] Cf. Van Bladel (2010: 77). The dates of al-Ṭabarī the physician (not to be confused with the historian al-Ṭabarī) are highly problematic (Thomas 2000). The tradition that the great physician Abū Bakr Muḥammad b. Zakariyyā' al-Rāzī (Rhazes, d. ca. 925) was his pupil is given strong credence from al-Rāzī's works. This would mean that al-Ṭabarī was born well after the traditional (but poorly supported) date, 781.

2. Al-Kindī (Alkindius, ca. 801–866), the first great natural philosopher in Islam, wrote works on astrology-astronomy, chemistry, logic, mathematics, music, medicine, physics, and psychology, among other topics. He was also knowledgeable about Indian thought, and personally copied a much-esteemed anonymous work of the late eighth century on the religious beliefs of the Indians.[52]

3. Muḥammad ibn Zakariyyā' al-Rāzī (Rhazes, d. ca. 925), a native of Rayy (near modern Tehran),[53] is generally considered to have been the greatest physician of Classical Arabic civilization. He wrote mainly on medicine, music, alchemy-chemistry, and philosophy. Philosophically he was a sceptic or atheist.[54] In medicine he belonged to the school of 'Alī ibn Sahl Rabban al-Ṭabarī and like him incorporated knowledge from both the Greek and the Indian traditions.[55] His medical works were still influential into early modern times.

4. Al-Battānī (Albategni, Albategnius, or Albatenius, d. 929), a Syrian from Ḥarrān, wrote works on astrology-astronomy and mathematics. His calculations and discoveries were influential as late as Copernicus.[56]

As this tradition grew, it came to be overwhelmingly dominated by Central Asian scholars and scientists. In fact, the most famous and greatest of the Classical Arabic scholars in the two centuries after the initiation of the Greek translations were largely from Central Asia. Besides the earlier scientists al-Khwārizmī (from Khwārizm) and 'Alī b. Sahl Rabban al-Ṭabarī (from Marw), mentioned above, there was al-Farghānī (Alfraganus, fl. 833–861, from Ferghana), al-Fārābī (Alfarabius or Avennasar, d. 950, from Fārāb[57]), Ibn Sīnā (Avicenna, 980–1037, from Afshana, near

[52] Fakhry (1983: 32). The book is unfortunately lost. On al-Kindī in general, see the study by Adamson (2007).

[53] Fakhry (1983: 97) puts Rayy in Khurasan (Central Asia), probably incorrectly, but in al-Rāzī's day the city was in any case not a part of Fārs 'Persia (proper)', and its importance was partly due to the fact that it was an important Western component of the Silk Road commercial system of Central Eurasia, q.v. Beckwith (2009).

[54] "Both al-Rāzī's atomism and his concept of the five eternal principles exhibit a striking similarity to the teaching of the Nyaya-Vaishishka [*sic*, for Nyāya-Vaiśeṣika—CIB] sect" (Fakhry 1983: 33 n. 127), which means it is similar to the views of the Buddhist Sarvāstivāda school as well. Cf. the citations by Fakhry (1983: 32–33) of other early Muslim works about Indian thought and their formative influence on al-Rāzī. It is also notable that his main opponents were nearly all from Balkh. Did he spend time there?

[55] Fakhry (1983: 33) and others comment on the evident Indian influence on his philosophy. This topic deserves serious attention.

[56] Nallino (1986), who also notes, "His excellent observations of lunar and solar eclipses were used by Dunthorne in 1749 to determine the secular acceleration of motion of the moon."

[57] Later known as Utrār.

Bukhara), al-Bīrūnī (973–ca. 1050, from Kāth, in Khwarizmia), al-Ghazālī (Algazel, 1058–1111, from Ṭūs in Khurasan), and many others.[58]

Of course, there were exceptions. For example, al-Kindī (Alchindus or Alkindus, d. ca. 870), an Arab from Kufa (in southern Iraq), was one of the first great Muslim natural philosophers. He founded a "Central Asian" school of Aristotelian Neoplatonism, in contrast to the Neoplatonic Aristotelianism of al-Fārābī and Avicenna, both of whom were native Central Asians.[59] Ibn al-Ḥaytham (Alhazen or Avennathan, ca. 965–1039), who was from Basra (his ethnicity is unknown), is generally considered to have been the greatest of the mathematicians and physical scientists of Classical Arabic civilization.[60]

## The Classical Arabic Recursive Argument Method

The first writer known so far who used the recursive argument method in Arabic works[61] was the Central Asian scientist and philosopher Ibn Sīnā, better known under the Andalusian pronunciation of his name, Avicenna. He was born in Afshana, near Bukhara, and was educated completely in Central Asia. His father, who was also a scholar, was from Balkh,[62] which had been the greatest Buddhist center of western Central Asia before Islam. Avicenna's autobiography does not make it absolutely clear, but it seems that he first studied at home with his father and "a teacher of the Qur'ān and a teacher of literature,"[63] probably first in Afshana and later certainly in Bukhara, while still a child.

[58] Despite the frequently repeated claim that these scholars were "Persians"—or, sometimes, "Turks"—none of them were Persians, either ethnolinguistically or politically, and few if any of them were Turks. They were Central Asians. The paucity of Persians among the great scientists of Islamic civilization seems to be due to most of Persia having remained Zoroastrian, and aloof, long enough to miss most of the great age of Classical Arabic civilization. The development and flourishing of Persian Islamic culture came only later.

[59] Adamson (2007: 14–15).

[60] Vernet (1986).

[61] On the extremely brief reference to the method in the *Encyclopaedia of Islam*, s.v. *shubha*, see this chapter, note 74. It is possible that some earlier scholar uses the method, but all the possibilities that have suggested themselves to me through my own research, as well as those suggested to me by others, have turned out to lack any actual examples of the recursive argument. If there are any earlier examples in Arabic, they are most likely to be found in theological (*kalām*) or legalistic (*fiqh*) works, especially, perhaps, commentaries, written by Central Asians. Scholars after Avicenna do use the method—see below on Fakhr al-Dīn al-Rāzī (who is often considered to have been primarily a theologian rather than a philosopher)—but some do not, perhaps for a reason. See the discussion of al-Ghazālī in chapter 7.

[62] Gohlman (1974: 16–17); Gutas (1988: 23); Goodman (2006: 11).

[63] Gutas (1988: 23); cf. Gohlman (1974: 18–19).

The recursive argument method was, or had been, widely known and used by Buddhists in Central Asia, who also practiced oral disputation (see chapter 4) in their colleges (see chapter 3), so it may be presumed from the outset that Avicenna did not learn it from written texts. It is unknown whether any translations of Buddhist scholastic texts were made into Classical Arabic before Avicenna's time. If any had indeed been translated, they could well have ended up in the library of the Samanids in Bukhara, which Avicenna was given permission to use by the ruler Nūḥ ibn Manṣūr, whom he had treated for an illness. Avicenna says in his autobiography:

> I was admitted to a building which had many rooms; in each room there were chests of books piled up on top of the other. In one of the rooms were books on the Arabic language and poetry, in another, on jurisprudence, and likewise in each room [were books on] a single science. So I looked through the catalogue of books by the ancients and asked for those I needed. I saw books whose names few people had encountered and which I had not seen before that time, nor have I seen since. I read those books.[64]

He adds that he mastered the contents of the books, noting that he had done this by the time he was eighteen years old.[65] Nevertheless, the presence of translations of Buddhist works is only a tantalizing possibility. Moreover, Avicenna seems not to refer explicitly to Buddhist or "Indian" ideas, though this no doubt bears further investigation.

However, in his autobiography Avicenna actually tells us explicitly how he learned the recursive argument method. He mentions that while a young student he studied "philosophy (*al-falsafa*), geometry (*al-handasa*) and the mathematics of India (*ḥisāb al-Hind*) [and algebra],"[66] and his father sent him to a grocer who used Indian mathematics, which Avicenna learned from him.[67] But he also studied Islamic law or jurisprudence (*fiqh*), in which he early became a skilled debater.[68] He tells us in his own words exactly what this entailed:

> I used to occupy myself with jurisprudence, attending frequently Ismā'īl az-Zāhid about it. I was one of the most skillful *questioners*,

[64] Translation by Gohlman (1974: 36–37), with a few minor emendations; cf. Gutas (1988: 29, 84).
[65] Gohlman (1974: 36–37); Gutas (1988: 84).
[66] Gohlman (1974: 20, 20 n. 2, 21); cf. Goodman (2006: 12).
[67] Gohlman (1974: 20–21); Gutas (1988: 24; 152 et seq.).
[68] Gohlman (1974: 20–21, 26–27); Gutas (1988: 25, 27); Goodman (2006: 12).

> having become familiar with the methods of *posing questions* and ways of *raising objections* to a *respondent*[69] in the manner customary with these people. Then I began to read the *Eisagoge* with an-Nātilī,[70] and when he mentioned to me the definition of "genus," viz. "that which is predicated of many things differing in species, in answer to [the *question*] 'What is it?'," I set about a verification of this definition in a manner such as he had never heard.[71]

According to the method Avicenna learned, using the terms in his own description, arguments had these main parts:

I. A Question
II. Responses to the Question
III. Objections to the Responses to the Question

In other words, according to Avicenna, the method consisted of an ARGUMENT, which was disputed by ARGUMENTS, which were disputed by ARGUMENTS. This is an exact, explicit description of the recursive argument method. It can be restated, using the terminology of the present book, as follows:

I. ARGUMENT (the MAIN ARGUMENT, QUESTION, or TOPIC)
II. SUBARGUMENTS$_1$ about the ARGUMENT
III. SUBARGUMENTS$_2$ about the SUBARGUMENTS$_1$ about the ARGUMENT

Avicenna thus tells us explicitly that he learned the recursive argument method, via oral disputation, from Ismāʿīl al-Zāhid, a legal scholar of the Ḥanafī school; that it was a traditional method used by him and "his people"; and that when he used the method to answer a question posed by his new philosophy teacher, al-Nātilī, he answered "in *a manner* such as he [al-Nātilī] had never heard." Avicenna's intention was "not to refer to the *subject matter* of jurisprudence, which is properly a non-philosophical discipline, but to its *method*" of dialectical disputation.[72] In

[69] I.e., Objections to the Responses to the Question.

[70] Avicenna earlier introduces al-Nātilī with the remark that he "claimed to be a philosopher" (Gutas 1988: 155; Gohlman 1974: 20–21).

[71] Text from Gohlman (1974: 20–23); translation of Gutas (1988: 155, his emphasis removed), my added emphasis. Note that his use of the method to answer the question "What is it?" is identical to the use of the method for the same question in the early argument quoted from text to text in early Central Asian Buddhist works, as discussed in chapter 4, though there the specific topic is the meaning of *Sarvāstivāda*.

[72] Gutas (1988: 155–156), emphasis added.

short, he learned a local intellectual tradition, understood its usefulness, and adopted it for his own works.[73]

In modern Islamicist scholarship, this method is mentioned extremely briefly, essentially as an afterthought. Rowson remarks that it is found in "theology and philosophy." He says, "In later scholastic treatises, positive arguments for a given view are often followed by a series of *shubah* counter-arguments by opponents, and their refutations."[74] This appears to be the same thing that Avicenna refers to in his autobiography.[75] His explanation of his training in it is explicitly presented as an aside to his comments on his first philosophy teacher, al-Nātilī, which are intended to explain that Avicenna had already learned the fundamentals of dialectics and logic from his training in legal disputation,[76] a subject he learned so well that he practiced it professionally for a time later in his life.

The recursive argument method is unusual as a literary form, to the point of being peculiar, so al-Nātilī's surprise at the method Avicenna used in his answer is perfectly understandable. Does this mean the method was not yet known outside of Central Asia? In the Middle Ages the method seems to appear suddenly outside its home culture: first in Tibetan, then in Arabic, and finally in Latin, and in each case is like nothing else that came before it in those literatures.[77] It has long been known that the Medieval Latin form of the argument has no classical antecedents; the same is true of the Arabic form. In the case of Tibetan, the recursive argument method was introduced via translations from Sanskrit (though it never caught on for native Tibetan works),[78] but the method used in

[73] It remains to be seen whether specialists in the literature of medieval Islamic jurisprudence can find examples of its use in written works of the Ḥanafī or other legal traditions.

[74] Rowson (2011) is an article I found essentially by chance, indicating the difficulty of finding anything about the topic in Islamicist literature. In Avicenna's *Metaphysics* the word *shubha* 'doubt' is used for a SUBARGUMENT$_1$ about the MAIN ARGUMENT, just as described. These extremely brief comments (plus a reference to a text that does not seem to use the method) are all that Rowson has to say about this meaning of *shubha*; most of his discussion in the article is about a completely unrelated usage (an 'illicit act' in a legal sense) in literature on Islamic law. His Arabic transcription has been regularized to follow the system used in this book.

[75] However, Islamicists need to look into this carefully.

[76] It is possible that Avicenna refers to it again in a letter to an anonymous disciple, in which he mentions "first raising doubts and then solving them" (Gutas 1988: 116).

[77] Latin Western Europeans were by far the most enthusiastic among these peoples in their adoption and adaptation of the recursive argument method, and used it in many fields of knowledge. In other cultures, even those scholars who did use the method did not use it as frequently as treatise-style arguments, and many scholars did not use it at all.

[78] See chapter 7.

Sanskrit texts is demonstrably adopted from earlier Central Asian texts, which were written in Gāndhārī Prakrit.

It is hardly conceivable that the highly distinctive recursive argument method could have been developed *de novo* twice, independently, in the very same region—Western Central Asia—and that it was used by the later Muslims in that same region in precisely the same way and for the same kinds of subjects as the earlier Buddhists, who had developed it themselves long before the coming of Islam and had used it over a very long period of time there.[79] The only problem is the lack of sources on its transmission from the Buddhists to the Muslims.[80]

It is necessary to digress for a moment on this point. The *madrasa* first appears in the Islamic world in Central Asia. Its special architectural design, legal endowment, educational functions, and concentration on theology or law, and so on, correspond to those of the earlier Buddhist *vihāra*, which was widespread in exactly the same region of Central Asia. The recursive argument method also first appears in the Islamic world in Central Asia, in works on theological metaphysics and natural philosophy by Avicenna, whose father was from Balkh, the former center of Buddhism in Central Asia; Avicenna says in his biography that he learned the highly unusual recursive argument method from a local teacher of Islamic law.

Some object to the "gap in the historical record,"[81] without which they are not willing to see transmission having taken place from the particular institutions and practices of the Central Asian Buddhists to the same exact institutions and practices of the Central Asian Muslims. That is, they demand an explicit statement in a written text so that they can believe.[82] This position necessitates supposing that when the Central Asian Buddhists converted to Islam—as written historical sources (supported by archaeology, art history, etc.) tell us very clearly they did—they *forgot* the functions of the buildings they were using, they *forgot* how to argue in their traditional highly distinctive manner, and

[79] The choice is simple inheritance or transmission versus miraculous independent reinvention, astounding coincidence, magic, and so on.

[80] The close interrelationships between Buddhists and Muslims in both cities and rural areas of Central Asia must have facilitated the exchange of ideas, but we know practically nothing about this. For the most recent, careful analysis of related topics, see Van Bladel (2010). For a tantalizing study of the Buddhist-Islamic interface via royal marriages in Ladakh and Baltistan in later times, see Halkias (2010).

[81] It must be pointed out that in ancient and medieval history, the gaps far outweigh the bits filled in with explicit accounts of anything. This seems not to be understood by many historians today.

[82] Their preferred alternative—belief in miracles and magic—hardly seems a good choice.

they *forgot* everything else they were doing according to their Buddhist traditions. Then, as Muslims, they or their descendants very creatively *invented* "new" architectural forms, endowments, and educational functions, and they *invented* an extremely distinctive way of arguing, which, miraculously, was absolutely identical to the one used by the same people in exactly the same places.[83] It is in fact exceedingly rare in medieval historiography that we have such a precise concatenation of data attesting to a historical event. We often have only a bare statement in a chronicle, which we must interpret with virtually no context. Such statements are practically meaningless by comparison with the rich data we have in the present instance.

On the Islamic college or *madrasa*, then, the accepted, traditional view of its early history is that it first appeared in what used to be called "Eastern Islam," or (misleadingly) "Eastern Iran," in both cases actually meaning Islamic Central Asia. As discussed in chapter 3, the first *madrasa* so far noticed in historical sources was founded in Bust in southern Central Asia (now southwestern Afghanistan) in 890, less than ninety years after the fall from power of Yaḥyā al-Barmakī and the other Barmakids in 803. Many more *madrasas* are mentioned as already existing a few decades later in Nishapur and other Central Asian cities.[84] By the early eleventh century, many *madrasas* existed in the area of Khuttal, a mostly rural part of Ṭukhāristān (ancient Bactria), where previously there had been many *vihāras*.[85] The *madrasa* is identical to the Central Asian *vihāra* in every respect except religious affiliation. All this is clear evidence of the

[83] I reject the wonderful, marvelous "coincidences" necessitated by this sort of pseudosceptical, pseudoscientific approach. In all fields of scholarship, including history, it is necessary once in a while to use commonsense logic.

[84] Bulliet (1972); cf. chapter 3. When the manuscript of this book was already in press I remembered an article by Richard Bulliet that I had read decades ago, found it online, and added this. Bulliet (1976: 145) quotes the thirteenth-century historian ʿAṭā Malik Juvainī's etymology of the name *Bukhārā*. Juvainī says it is from a word *bukhar*, which in the language of the "Magians"—i.e., Persian—means "centre of learning." Most significantly, he says, "This word closely resembles a word in the language of the Uighur and Khitayan idolaters, who call their places of worship, which are idol-temples, *bukhar*." As Bulliet notes, since Juvainī is clearly equating *bukhar* with *vihāra*, "the implication of this report is that old Buddhist monasteries in Iranian territory retained a reputation as educational centres for centuries after they had lost their purely religious identification." If Bulliet is right, this could well reflect the history of the conversion of the Buddhists—including the Barmakids, who retained hereditary control over the Nawbahār of Balkh and its endowments—to Islam, and along with them the conversion to Islam of related institutions in the region. In that case, the early appearance of so many *madrasas* in Nishapur (Bulliet 1972) is explained, and the only novelty was the change of name from *bahār* (*vihāra*) or *nawbahār* (q.v. Bulliet 1976) to *madrasa*. It also would contribute to our knowledge of the transmission of the recursive scholastic argument to Islam.

[85] Litvinskij (1985).

conversion of the population to Islam. As in other cases, the institutions' religious affiliations changed, but not the architectural forms, functions, or legal status—described in Arabic sources as a *waqf* 'pious endowment', the same term used for the endowment of a *madrasa*[86]—which remained essentially unchanged. This cannot be accidental. The equivalence must be due to the direct transmission of the *vihāra* tradition by live people. That is, the institutions were converted to Islam along with the people.[87] It is now known that by Ibn Sīnā's lifetime the *madrasa* was ubiquitous in Central Asia.[88] Its importance as the lone college institution during the great age of Islamic civilization cannot be overemphasized, though the use to which it was put deserves very careful study.

Makdisi illuminates the great importance of public oral disputation in the medieval *madrasas* and in the medieval European colleges and universities. The same can be said about the *vihāras*. As Makdisi shows for medieval Islam and medieval Latin Europe, oral teaching and disputation were closely connected to literary teaching and disputation, both of which were very energetically practiced in both *madrasas* and colleges.[89] Makdisi accordingly argues that the recursive argument method (the "scholastic method") is an oral form of disputation that was transmitted to Western Europe through live oral disputation in the *madrasa*, which institution he contends was transmitted, whole, to the Latin West. Since the formal method of oral disputation that Avicenna learned in his study of jurisprudence is identical in structure to the literary recursive argument method he uses in his *Metaphysics* and *De anima*, it must be the source of the Latin method too, essentially as argued by Makdisi.[90]

[86] Van Bladel (2010: 64–65) translates an early account of the Nawbahār that explicitly mentions the *waqf* properties controlled by the Barmakids; see chapter 3 and above in this chapter.

[87] See de la Vaissière (2009) on the continuity of tradition from the Buddhist to the Islamic periods in caravanserais and similar institutions.

[88] Unfortunately, Avicenna does not tell us *where*, exactly, he learned disputation from the Ḥanafī jurist Ismā'īl al-Zāhid. Though it was certainly in Bukhara, and *madrasas* no doubt existed there by that time, Avicenna was not a poor scholar. As an aristocrat, a son of a high-ranking official of the Samanid government, he was educated by tutors. But that does not necessarily mean he learned something different from what was taught in the *madrasas*. Was it a specifically local method of oral disputation, or a Ḥanafī method? Central Asia was predominantly Ḥanafī.

[89] Makdisi (1981).

[90] Cf. Grant (1996). Note that formal oral *debate* was not the same as the fully developed recursive oral *disputation*. Traditional formal debate was resolutely Q : A (or B : A) in structure, as attested by numerous literary examples of debates—and even handbooks on debate—in Greek, Sanskrit, Classical Arabic, and Medieval Latin. I have not found any handbooks or descriptions of the recursive argument method in ancient or medieval sources (with the partial exception of the Indian method, q.v. chapter 7), except for the remarkable description given by Avicenna of the method he learned as a child, but the procedure of the oral version of the recursive argument

The three traditions had teachers and colleges in which knowledge was conveyed largely orally to the students, who in at least some cases recorded the disputation. In Islamic Central Asia the colleges were *madrasas*, which largely continued the teaching and research program of the *vihāras*. The basis of the formal curriculum was jurisprudence (*fiqh*, the study of Islamic law, corresponding roughly to the study of canon law in Christianity), though college education as a whole was not restricted exclusively to that subject.

In the Buddhist tradition, the vast majority of scholarship, including the texts in which the recursive argument method is found, belong not to the *vinaya* (monastic rule), but to other subcategories of Buddhist teachings. The Sarvāstivāda school's major scholarly texts are on *abhidharma* 'clarification of *dharma*'—referring to both *the* Dharma 'the Law' and *dharmas*, 'momentary constituents of phenomena'.

Although in the Islamic tradition religious law became an important subject of study in itself, for most students it was simply a prerequisite to further study. This is clearly the case for Avicenna, whose early education consisted in large part of Islamic studies (the Koran and *fiqh*), despite his interest in other things. In both the Buddhist and the Islamic periods of Central Asian culture, education was based on the fundamental teachings of the respective religions, but the great literary works relating to religion were not written on religious law *per se* (i.e., *vinaya* in Buddhism, *fiqh* in Islam, canon law in Christianity), but on theology and metaphysics. These were the topics of greatest interest to the leading intellectuals of each culture. Although philosophy and science were pursued by a tiny minority of scholars in the Islamic tradition, some of them were among the most famous in history.

In short, the recursive argument method was not transmitted by sea, by land, or in any other commercial-like transaction. It did not "go" anywhere at all in its transmission, which took place in situ in Central Asia as a part of Central Asians' conversion to Islam. The recursive argument method was thus transmitted from Central Asian Buddhism to Central Asian Islam along with an extensive package of developed intellectual culture, including many other elements, of which the most important was the institution of the college—the *vihāra*, or *madrasa*, with its functions, methods, and distinctive physical architectural form.

---

method, or *quaestiones disputatae* 'disputed questions' method, is referred to in Medieval Latin works. See the valuable papers in Bazàn et al. (1985); cf. Little and Pelster (1934).

As discussed in chapter 2, the recursive argument method is used most saliently in great summas, comprehensive works by major scholars intended in part for students. In the Classical Arabic tradition it seems not to have been used in medicine or the other "practical" sciences,[91] but mainly in psychology, theology, and what Avicenna, following Aristotle, considered to be the most important science, metaphysics, which in such heavily religious cultures necessarily included much that may be classified as theology—a point not missed by medieval scholars.

It is thus not surprising that within his summa, the *Kitāb al-shifā'* 'The Book of the Healing', Avicenna uses the recursive argument method in the *De anima* 'On the Soul', or 'Psychology', and the *Metaphysics*, but apparently not in the medical part. In Example 5.1 below, note his use of the recursive method at the macro level (i.e., full recursive argument, each of which constitutes in turn one of the major arguments in the book) and at lower levels (full recursive arguments *within* individual subarguments[92] of a full recursive argument).[93] For simpler, clearer recursive arguments by Avicenna, see Example 6.1 from the *De anima* and Example 6.3 from the *Metaphysics*, in the next chapter.

### *Example 5.1*

**Ibn Sīnā (Avicenna)**
***Kitāb al-Shifā', Al-Ilāhiyyāt* ('The Metaphysics')**

Book 6, Chapter 5

Chapter [Five][94]

[I. Main Argument (Topic or Question)]

On establishing purpose . . .

[91] Fransen (1985: 250ff.) discusses *quaestiones disputatae* in medieval law, but the examples given do not seem to be recursive arguments. Further study is therefore needed.

[92] The terms *shubha* 'doubt, obscurity, uncertainty, specious argument' and its functional equivalent *shakk* 'doubt'—which are freely interchangeable in the texts and traditionally translated as 'objection'—are not always mentioned explicitly.

[93] For further examples of Avicenna's method, see chapter 6.

[94] Arabic text: Marmura (2005: 194, 220–234), corresponding to the Cairo ed. (Avicenna 1960, 2: 283–298). For complete translations of the argument analyzed here, see Marmura 2005: 220–234 (English) and Anawati 1985: 36–47 (French). Though unnoted in the published text, the last part of the chapter is actually the ending of book VI as a whole (Marmura 2005: 234–235); cf. the Cairo ed. (Avicenna 1960, 2: 298–300) and the French translation (Avicenna 1985: 47–48).

[II. AUTHOR'S VIEW]

We say (*naqūlū*): It has become clear, from what we have previously stated, that . . .

[III. SUBARGUMENTS$_1$]

[1] But it has not become clear that . . .
[2] Then someone may say (*thumma liqā'il an naqūl*) . . .
[3] Then someone may say . . .
[4] After solving this [the third] argument (*al-shubha*) we should also debate:

[4.1] whether the end and the good are one thing or are different,
[4.2] and also what the difference is between magnanimity and benevolence.

[IV. SUBARGUMENTS$_2$]

[1] As for the first argument (*al-shakk al-awwal*) . . . , we will solve it.

[SUBORDINATE ARGUMENT]

[I. MAIN ARGUMENT (TOPIC or QUESTION)] [The first argument]
[II. AUTHOR'S VIEW]

We say: As for . . . [Three pages of detailed discussion follow.[95]]

Therefore, having established these premises, we say:

[III. SUBARGUMENTS$_1$]

[1] The statement . . . is a false statement.[96]
[2] Moreover, the statement . . . is a false statement.

[IV. SUBARGUMENTS$_2$]

[1] As for the first (*ammā al-awwal*) [it is false because] . . .
[2] As for the second (*ammā al-thānī*) [it is false because] . . .

[2] As for the argument (*al-shakk*) that follows it [i.e., the second],

. . . [Three pages of refutation follow.]

[95] Such approximations, intended to give the reader a rough idea of the extent of some of the longer arguments, are based on Marmura's text.

[96] In this particular SUBARGUMENT$_1$ section, Avicenna declares that the quoted statements are false, and then in the SUBARGUMENTS$_2$ he disproves them. Kamalaśīla does the same thing in some of his SUBARGUMENTS$_1$; see Keira (2004).

[3] As for the argument (*al-shakk*) that follows it [i.e., the third], . . .

[Two pages of refutation follow.]

[4] As for the discussion [of the problems, [4.1] and [4.2], we said should follow] after this, it is revealed from what we say that:

[4.1. FIRST SUBORDINATE ARGUMENT]

[I. MAIN ARGUMENT (TOPIC OR QUESTION)] [As for the first, the problem of] the end that exists in the act of the actor, it may be divided into two parts:

[II. SUBARGUMENTS$_1$]

[1] an end that is a form or an accident in a patient, receptive of an action[97]

[2] an end that is not at all a form or an accident in a patient receptive [of the action], and is thus inescapably in the agent. . . . [98]

[III. AUTHOR'S VIEW]

It appears that . . . [One paragraph.] Now that this has been established, we say:

[IV. SUBARGUMENTS$_2$]

[1] As for the first part [i.e., the first SUBARGUMENT$_1$], the end is connected to many things . . .

[2] As for the end according to the second part [i.e., the second SUBARGUMENT$_1$], . . .

[4.2. SECOND SUBORDINATE ARGUMENT]
As for [the second of the two problems we said should follow, namely,] the case of magnanimity and benevolence, it must be known that . . . [99]

Fakhr al-Dīn al-Rāzī (1149/1150–1209/1210), an important later scholar and philosophically inclined theologian who criticizes Avicenna as well as philosophy per se in general,[100] uses the recursive method in a

[97] Translation of Marmura (2005: 229–230).

[98] Translation of Marmura (2005: 230).

[99] This subordinate argument within the fourth SUBARGUMENT$_2$ is presented treatise-style, unlike the previous one.

[100] Nasr (2006: 38–39, 113–114, 127–128).

creative way in his *Kitāb al-arbaʿīn fī uṣūl al-dīn* 'The Book of the Forty [Questions] on Theology'. Note his use of numbering in both lists of arguments, among other features, in Example 5.2.

### *Example 5.2*

**Fakhr al-Dīn al-Rāzī**
***Kitāb al-arbaʿīn fī uṣūl al-dīn***

Question 24[101]

[I. Main Argument (Question or Topic)]

On stating that the Exalted One is the Willer[102] of all creation. The Muʿtazilite school: [God's] will is in harmony with [His] command. Everything that God commands He has willed it, and everything that He rejects, He has disliked it. Our school: [God's] will is in harmony with knowledge. Everything that is known to happen is intended to happen. Everything that is known not to happen . . .

[II. Author's View Argument]

On this we have two views (*wujhāni* [sg. *wujha* 'viewpoint']) . . .

[III. $\text{Subarguments}_1$]

As for the Muʿtazilites, they advance the [following] views (A. *wujūh*) to support their doctrine:

[1] The first doubtful argument (*al-shubha* 'obscurity; doubt; specious argument'): . . .
[2] The second . . .
[3] The third . . .
[4] The fourth . . .
[5] The fifth . . . This is refuted by reason and listening.

[5.1] By reason: . . .
[5.2] By listening: . . . [quotations from the Koran]

[6] The sixth . . .

[101] Arabic: Al-Mūsawī: 244–246.

[102] A. *murīd*. The usual senses of this word ('religious aspirant; kind', etc.) do not work here in the context of the discussion of God's will (*irāda*) and his willing all creation into existence. Clearly, al-Rāzī intends it to have the general etymological sense of the agentive nominal form of the root *rwd*, the base of *irāda* 'will', which word occurs throughout the argument.

[IV. SUBARGUMENTS$_2$]

[1] Reply to (*al-jawāb ʿan*) the first doubtful argument: . . .
[2] To (*ʿan*) the second . . .
[3] To the third . . .
[4] To the fourth . . .
[5] To the fifth . . .
[6] To the sixth . . .

As already noted, one of the most striking things about Fakhr al-Dīn's form of the method is his explicit numbering of all of the subarguments in both lists, but his other innovations, in particular the organization of the beginning section, are of interest as well and deserve attention by specialists in this literature. He is clearly innovative in his use of the recursive argument, much like William of Ockham a century later.

# — Chapter Six —
# Transmission to Medieval Western Europe

The appearance of the recursive argument method in Latin texts was preceded by more than a century in which Classical Arabic learning was increasingly translated and introduced to the Medieval Latin world. A trickle of translations of Arabic scholarly books into Latin had already begun to appear in Italy and Spain by the mid-eleventh century,[1] but none of the works known to have been translated at that time seem to use the Arabic version of the recursive argument method.

Intellectual culture in Western Europe remained very conservative until the second half of the twelfth century. Despite the undoubtedly important advances in theology that had lately been made by Anselm of Canterbury, Peter Abelard, and others, their great authorities were mainly St. Augustine, St. Jerome, other Church fathers of late antiquity, and above all Scripture itself. Arguments in texts from the period are structured according to either treatise or dialogue format. There are no examples of recursive arguments in Latin from this early period.

The recursive argument method first appears in Western Europe in Avicenna's *De anima* 'On the Soul' or 'Psychology', the subtitle of which in Latin is 'The Six (Books) on Nature'. It is the Latin translation of Avicenna's *al-Nafs*, part of his great summa, the *Kitāb al-shifā'* or 'The Book of the Healing'. The translation, done in Toledo by "the Jewish philosopher Avendauth [Ibn Dā'ūd]" and "Dominicus the Archdeacon", is dedicated to Archbishop John of Toledo (r. 1152–1166). This explicit information, given in the preface, establishes the period when the work's Latin translation was completed; it seems likely to have been begun under John's predecessor, Archbishop Raymond (r. 1125–1152).[2] Consider the sample argument outlined in Example 6.1.

[1] See d'Alverny (1982).

[2] d'Alverny (1952: 356). For details on the translations, translators, and chronology, see appendix A.

## *Example 6.1*

### Avicenna (Ibn Sīnā)
### *De Anima* (*On the Soul*, or *Psychology*)

### Part V, Chapter 7[3]

Chapter in which credible views (*sententiae*) of the ancients on the soul and its actions are listed, and whether there is one [soul] or many [souls], and establishing the truth of the matter about it.[4]

[I. Main Argument (Question, Topic)]

On the soul and its functions: whether there are one or many.[5]

[II. Subarguments$_1$][6]

There are different views (*sententiae*) on the essence of the soul and its actions. . . .[7]

[1] Those of the first school (*secta*) say that the soul is not one, but many, and that the soul which is one in a body is a collection of souls . . .[8]

[2] Some among them, on the other hand, say that the soul is rational by itself . . .

[3] But the authors of views reasoning on the basis of recollection (*reminiscendo*) say . . .[9]

[III. Author's View Argument]

[3] Latin: Van Riet (1968: 154–174); French translation: Bakoš (1956, 2: 177). The three Subarguments$_1$ begin on page 155, line 44; page 155, line 50; and page 156, line 65. The three Subarguments$_2$ begin on page 167, line 15; page 167, line 23; and page 169, line 51. The final Author's View section follows, from pages 169 to 174.

[4] Lat. "*Capitulum in quo enumerantur sententiae antiquorum probabiles de anima et eius actionibus et an sit una aut multae et certificare et stabilire veritatem rei in hoc.*"

[5] The Main Argument is in the chapter title, q.v. the preceding note.

[6] This section is introduced by the comment "*Sententiae de essentia animae et eius actionibus sunt diversae*," etc.

[7] Avicenna here gives a very brief list of the different views, perhaps intended to function as a sort of table of contents.

[8] Van Riet (1968: 167 n. 15) says this argument begins on page 155, line 44, of his edition, but this is clearly an error.

[9] The end of this section is marked formally by Avicenna's comment, "These are the notable views on the soul. . . . After [the Author's View argument section, which follows this comment] we will return to solve the subarguments (*quaestiunculae* 'little arguments') introduced [above]."

We say, therefore (*Dicemus igitur*) . . . [10]

[IV. SUBARGUMENTS$_2$][11]

Since we have now shown the certitude of this [Avicenna's own] view, we need to solve the above-mentioned questions (*quaestiones*).

[1] Because the soul is one essentially it should not therefore . . .

[2] But the theory (*ratio*) of those who argue that the soul is sentient by itself . . .

[3] As for the theories of the advocates [of the theory based on] reminiscence, we have destroyed them in the *Metaphysics*.[12]

[V. AUTHOR'S VIEW ARGUMENT]

The theory of those who divide up the soul, moreover, assumes false premises. They say . . . [13]

Avicenna's *De anima* was an important text in medieval Western Europe. Part 3, *De visu* 'Optics', a thorough analysis of the different major theories on vision and light, was one of the most influential of his works. It is complex and technical—mathematics, physics, and physiology are some of the relevant scientific fields comprised in his discussion—and is therefore not presented here.[14] It must be stressed that this highly influential text for medieval European science, which continued in this respect the Classical Arabic scientists' interest in optics, was largely written in the novel recursive argument method. Some of its arguments are complex and require the reader to follow their structure in order to grasp

[10] This argument takes up ten pages in Van Riet's edition. Eliminating the critical apparatus would make the text alone about four pages long—still substantial.

[11] Latin: "*Postquam autem iam monstravimus certitudinem huius, debemus solvere quaestiones praedictas*." Latin *quaestiones* here translates Arabic *shubah* (pl. of *shubha*), 'doubts, difficulties'. The same argument (*De anima* V, 7, line 85) has *quaestiuncula* 'little question' (cf. *De anima* III, 7, lines 85, 88) corresponding to the original Arabic *shubha* 'doubt, difficulty' (Avicenna 1968: 157), while in the *Metaphysics* (VI, 5) argument analyzed below, *shakk* 'doubt, difficulty' (an Arabic synonym of *shubha*), is regularly used, in the same meaning, for the same purpose, but for both citations in the *Metaphysics*, the Latin translates *shakk* as *dubitatio* (Avicenna 1980: 332, 336). Latin *quaestiuncula*, which is often used instead, is the equivalent of my term 'subargument'.

[12] The position of the third SUBARGUMENT$_2$ is signaled by this sentence, but no actual argument is given; Avicenna refers the reader to the *Divina doctrina* 'Metaphysics' section of *Kitāb al-shifā'*. However, this reference may be a mistake; see the discussion by Van Riet (1968: 169 n. 52), who, however, does not say where such a discussion actually occurs in any of Avicenna's works. Bakoš (1956, 2: 230 n. 686) says, "C'est la Logique."

[13] About five pages of text (including the apparatus) in Van Riet's edition.

[14] See Van Riet (1968: 169ff.); for medieval optics, see Lindberg (1976).

them. No one could have read and understood the *De anima* without also understanding the systematic, rigorous method that suffuses the text: the recursive argument method. Incredibly, it appears that there is no translation of *De visu* from Latin into a modern European language.

The next earliest example of the recursive argument method in Latin is in the translation of Avicenna's *Metaphysics*. This part of *al-Shifā'* was translated from Arabic by several scholars, among whom the most consistently named is Dominicus Gundisalvi, archdeacon of Toledo, who worked with the Jewish philosopher Avendauth as well as with a Mozarab known as Master John, both of whom knew Arabic. The work was apparently completed in the third quarter of the twelfth century, shortly after the translation of *De anima*.[15]

Compare Avicenna's use of the recursive argument method in the example taken from the Latin translation of the *Metaphysics* and outlined in Example 6.2.[16] Another, much shorter, example from the same work is given in Example 6.3; the translation in this instance is based on both the Arabic and the Latin texts, which are extremely close.[17]

### *Example 6.2*

**Avicenna (Ibn Sīnā)**
***Liber de Philosophia Prima, sive Scientia Divina***[18]

Book 6, Chapter 5[19]

The Sixth Book

And in it are five chapters.

Chapter [Five]

[I. Main Argument (Question, Topic)]

On establishing purpose . . .

[15] See appendix A.

[16] For the Arabic original of this text, see Example 4.1a.1.

[17] The frequently repeated claim to the effect that the Latin translations were practically unintelligible is not supported by the texts. Of course, there are always problems with manuscript copies, and the quality of translation was certainly not even (nor is it today, by any means), but even taking that into consideration the translations are on the whole rather clear and close in meaning to the originals. See, similarly, the comments of Grant (2004: 166, 169; 2007: 138–140).

[18] This translates the Latin translation of the Arabic original, which is translated in the previous chapter, q.v. for several notes relevant here also.

[19] Latin text: Van Riet (1980: 326–346).

[II. AUTHOR'S VIEW ARGUMENT]

I say (*dico*): . . .

[III. SUBARGUMENTS$_1$][20]

[1] However, it is not yet accepted that . . .
[2] Again, someone may say (*Potest etiam aliquis dicere*) . . .
[3] And again, someone may say (*Et potest etiam aliquis dicere*): . . .

[4] Moreover, what should be asked after solving this [the third] question[21] [*quaestio*], is:

[4.1] whether the end and the good . . .
[4.2] also what the difference is between magnanimity and benevolence.

[IV. SUBARGUMENTS$_2$]

[1] So, [as for] the first question (*Primam igitur quaestionem*), . . . I will solve it.

[SUB-SUBORDINATE ARGUMENT, SCHOLASTIC-METHOD STYLE]

[I. MAIN ARGUMENT (TOPIC OR QUESTION)]
[II. AUTHOR'S VIEW]

And I say (*dico*): As for . . . [four pages of argument follow]. Accordingly, after having established all these premises [*Postquam autem constant hae omnes propositiones*], then:

[III. SUBARGUMENTS$_1$]

[1] The statement . . . is false.
[2] The statement . . . is false.

[IV. SUBARGUMENTS$_2$]

[1] First, that (*Primo, quod*) . . . [refutation of [1]]
[2] Second, that (*Secundo, quod*) . . . [refutation of [2]]

[20] The 'objections' or 'questions' (L. *objectiones, quaestiones*, translating A. *al-shukūk* or *al-shubhāt* 'doubts, obscurities, uncertainties, specious arguments') are explicitly mentioned as such only below.

[21] L. *post solutionem huius quaestionis* 'after the solution of this question'.

[2] The argument (*Dubitatio*)[22] [i.e., the second] that follows . . .

[four page refutation]

[3] But the argument (*Sed dubitatio*) [i.e., the third] that follows . . .

[lengthy refutation]

[4] As for [the two problems] that should be discussed after this . . .

[they are the following arguments]:

[4.1. First sub-subordinate argument, recursive method][23]

[I. Main Argument (Topic or Question)]

An end that occurs from the action of an agent.

[II. Subarguments$_1$]

[1] An end that is a form or intention in a patient receptive to action.
[2] An end that is not a form or intention . . . in any way.

[III. Author's View]

It appears that [*videtur quod*] . . . [one paragraph]. After, therefore, this has been established, I say that (*dico quod*) . . .

[IV. Subarguments$_2$]

[1] As for the first part, the end is connected to many things . . .
[2] As for the end according to the second part, verily . . .

[4.2. Second subordinate argument, treatise method]

As for investigating [the second of the two problems we said should follow, on]: the disposition of magnanimity and benevolence . . . It must be known that . . .

[22] At this point in the text, Arabic *al-shakk* is translated as *dubitatio* 'doubt'; after Avicenna's digression, *al-shakk* is translated as *quaestio* 'question'.

[23] This subordinate subargument is organized formally as a recursive argument.

## *Example 6.3*

### Avicenna (Ibn Sīnā)
### *Liber de Philosophia Prima, sive Scientia Divina*[24]

Book 5, Chapter 6

Chapter [Six][25]

On definition and verification of a *differentia*.[26]

[I. Main Argument (Question, Topic)]

It is also necessary to debate and make known the status of a *differentia*. We say: A *differentia* in reality is not like reason and sensation. . . . [one page of discussion][27]

[II. Subarguments$_1$][28]

[1] Then, an objection [lit., '(one) among the doubts (A. *al-shukūk*'; L.: *objectiones*)] that can be raised against this argument (A. *al-kalām*), or rather on the existence of the nature of a *differentia*, is[29] . . . [one paragraph]

[III. Author's View Argument]

What must be known in order to solve [L. *per quae solvitur*] this objection [A. *al-shakk* 'doubt'; L. *quaestio* 'question'] is that . . . [two and a half pages of discussion]

[IV. Subarguments$_2$]

[1] Let us return now to the premises [A. *muqaddimāt*; L. *propositiones*] of the objection [A. *al-shakk*; L. *quaestio*].[30] We say:

[24] This translates the Latin translation of the Arabic original, which is translated in the previous chapter, q.v. for several notes relevant here too.

[25] Arabic: Marmura 2005: 175–180; Latin: Van Riet 1980: 278–285; modern translations: Anawati 1960: 257–261 (French), Marmura 2005: 175–180 (English).

[26] The Latin translation has "*Capitulum de differentia et eius certitudine*."

[27] The Main Argument section could perhaps be analyzed as another Author's View section, but it seems best to be treated simply as an explanation or gloss on the meaning of a *differentia*.

[28] As is obvious, there is only one Subargument$_1$ in this argument. In this case, it represents a view opposed to the author's, so there is a Subargument$_2$, but if the lone Subargument$_1$ were one with which the author agreed, there would normally be no Subargument$_2$.

[29] Lit., 'is what I say'.

[30] Lit., 'the premises that are in the objection'.

[1.1] As for the premise that says the *differentia* is . . . it is conceded. . . . [brief discussion]

[1.2] As the other [premise], which says . . . , it is a false statement.

. . . [extended refutation]

Late twelfth- to early thirteenth-century Western Europe thus saw the more or less simultaneous appearance of the college, the recursive argument method, and translations of independent Greek and Classical Arabic scientific-philosophical works[31] as well as translations of many of Aristotle's works from Greek and from Arabic,[32] along with Arabic commentaries on them.[33]

Traveling clerics, French prelates posted in Spanish cities, and others kept scholars in France, England, and elsewhere in Europe in constant contact with the scholars working on translation of Arabic texts in Spain. Translations of Aristotle, Avicenna, and related works seem to have been in circulation in Paris within a decade or two after their translation. Because the translators focused on the one hand on works by classical Greek authors, especially Aristotle and his Arabic commentators, and, on the other, on Arabic scientific works, including the magisterial works of Avicenna on medicine and natural philosophy, the translations instantly acquired extremely high prestige in Europe. Western Europeans welcomed with open arms what became a flood of literature by philosopher scientists with the exotic Latin names Alfarabius, Algorithmus, Alhazen, Alkindius, Avicenna, Averroës, and many others. The result was the "intellectual revolution"[34] of the twelfth and thirteenth centuries in Medieval Latin Europe.

The newly translated texts became so popular so quickly that study of many of them by students of the University of Paris was banned in 1210 by a Church decree.[35] A similar decree issued in 1215, also considered

[31] Following slightly earlier translations of Arabic works on Indian mathematics.

[32] Some of Aristotle's works were translated more than once, from Greek and from Arabic, and in many cases older translations influenced the newer ones. See Grant (1974; 2004: 165–169).

[33] Cf. Grant (1996: 32).

[34] Southern (1995: 31); cf. Fidora (2003: 10).

[35] Grant (2007: 143ff.). In some modern accounts one reads that the Church *successfully* censored the new translations, at least in Paris. To which is invariably added, as if in proof of the claim, that edicts proscribing this or that text or position were repeatedly proclaimed, in such and such years, showing the success of the Church's putative repression. This is incorrect. It is commonplace knowledge in historiography that repeated proclamation of laws against something or other is *direct evidence* that the something was still going on essentially unchanged. In the present case, it is known that these edicts, however much they were ever actually in effect, were definitely no longer in force by 1255 (Grant 2004: 176ff.), and that the Church was actually the principal support of science in medieval Europe (Lindberg 2007: 225ff.).

to be the charter of the university, was promulgated by Robert of Curzon (Robert de Courçon,[36] d. 1218), an English cleric who studied and taught in Paris, but that decree was apparently also ignored. By 1255 all of Aristotle's works were being taught at the University of Paris. The new translations were officially approved (with the exception of a few specific arguments considered heretical), and were assigned as the new "liberal arts" curriculum—most of which consisted of logic and "natural philosophy"—that was required of all bachelor's level university students in Western Europe.[37]

The argument given in EXAMPLE 6.4, from Robert of Curzon's *De usura* ('On Usury') is the earliest known example of the recursive argument method in an original authored (i.e., not translated) text composed in Latin.[38] The Latin text has been edited and translated into French as an independent work, but it actually constitutes a part of Robert's *Summa*, which is now dated to ca. 1208–1215.[39]

The very first argument in *De usura* is a full recursive argument. Because the view opposed by the author in this argument is the *pro* view, not the *contra* view, he omits replies to all of the SUBARGUMENTS$_1$ that he gives *contra* the proposition in the MAIN ARGUMENT (the TOPIC, or QUESTION). He also omits a reply to the third view, though he evidently opposes it. There are a few other recursive arguments in this little work,[40] but most are not full arguments or they are not structured as recursive arguments at all. In including recursive arguments, treatise arguments, and variations of both, all in the same work, Robert's usage is reminiscent of Avicenna's practice in his *Metaphysics*. The extent to which Robert uses the recursive argument method further in his summa remains to be shown.

[36] "L'Anglais Robert de Courçon ou de Curchun (on écrit son nom de dix autres manières), chanoine de Noyon, était alors [ca. 1202] à Paris, où l'on suppose qu'il résidait depuis l'année 1195" (Hauréau 1890: 179). Robert presided over a council there in 1213, and died in a crusade against the Saracens in 1218. Cf. Lefèvre (1902: xv–xvi). He was English, so I follow Rashdall's (1936: 309ff.) usage, 'Robert Curzon', but include the 'of' as in the French form.

[37] Grant (2004: 176–177). For a thorough discussion, see Grant (2004: 169ff.). On the crucial importance of the universities, and the major changes in the medieval education system and curriculum after the translations, see Grant (1996: 36, 42ff.).

[38] Grabmann (1909–1911, 2: 495) notes that the full recursive scholastic method is first used by Robert, citing the same argument presented here; he analyzes the argument's parts as: "Fragestellung, Argumente, Gegenargumente (contra), Lösung (corpus articuli), Kritik der für die entgegengesetzte Meinung angenommenen Argumente." Note that the views the author opposes in this example are actually the "Argumente" (i.e., the *pro* view), not the "Gegenargumente" (the *contra* view). For the *Sentences* of Peter of Poitiers and the work's problematic dates, see appendix B.

[39] Macy (2009), q.v. on manuscripts and studies of Robert and his summa. Lefèvre (1902: xvi) dates Robert's *Summa* to 1202 (cf. Hauréau 1890: 179), but the information given by Macy indicates it should be dated approximately a decade later.

[40] E.g., Lefèvre (1902: 21–25, 39–43).

The originality of Robert's summa lies to a large extent in its completion, revision, and updating of the unfinished *Quaestiones* of Peter the Chanter. Just as Peter of Poitiers or one of his students must have revised his own *Sententiae* in various ways (including introduction of the recursive argument method) after the date of its dedication, Robert of Curzon took Peter the Chanter's work, which is written in the early eleventh-century version of the ancient nonrecursive *sententiae* or *quaestiones* method, and rewrote it, updating the work to include examples of the new recursive argument method.[41] These works adopted the new recursive method *and adapted the traditional European methods to accomodate it*, thus producing the Medieval Latin recursive argument method.

Robert's early life is obscure. He died in Egypt in 1218[42] during the Fifth Crusade. Had he been sent because of his knowledge of Arabic or the Holy Land? His references to "Saracens" (Muslims) in this text sound as if he had personal knowledge of them.

## *Example 6.4*

### **Robert of Curzon**

### *Summa*

### *De usura* (On Usury)[43]

[I. Main Argument (Topic or Question)][44]

The question of usury: whether in certain cases usury is admissible.

[II. Subarguments$_1$]

[a. *Pro*]

[41] Baldwin (1970: 25). See also chapter 2, and especially appendix B.

[42] Macy (2009) gives 1219 without comment.

[43] Lefèvre (1902: 3–7 [Latin], 2–6 [French]).

[44] *Quaestio de usura* 'the question of usury' is the marginal gloss title of the first *quaestio* 'question, argument' in *De usura*. It is not clear whether the question's title is original or has been provided by the editor. The marginal gloss adds also, "Where from the outset that which is 'usury' is defined (*Ubi in primis determinatur quid sit usura*)." The placement of this gloss beside the first paragraph may reflect the manuscript (which I have not seen) or it may be Lefèvre's work. It clearly belongs with the first and second paragraphs of the text as a whole, not with the first *quaestio* 'question, argument' proper, which in his edition begins with the third paragraph. The first paragraph gives a little introduction, "Here begins the topic of usury (*Post haec agendum est de usura*). We will see, therefore, what is usury, whether in certain cases it may be acceptable, what kinds of it there are, what punishment is to be inflicted on usurers . . . "

[1] That[45] (in certain cases usury is admissible) would appear to result from Ambrose (Chapter XIV, *quaestio* IV, *Ab illo*):[46] "You may exact usury from someone you have the intention of hurting, such as a Saracen."

[2] Again, in *Deuteronomy*,[47] "You do not charge your brother interest, but (only) others." Therefore, one may charge interest to someone who is not your brother.

[3] Again, the *Lex Justiniana*[48] permits one percent interest, and the lord Pope has allowed it in fact for lenders. . . .

[b. *Contra*]

[4] But the contrary view[49] has more valid reasons that cannot be resisted, such as that in Ezechiel: "You do not accept usury, nor any overpayment." And in Exodus . . . And Augustine . . .

[5] Again, in *Leviticus* . . . [50]

[6] Again, in *Exodus* . . . [51]

[7] And in the New Testament, too . . . [52] where the Lord says, "Give to each other, without hoping for anything in return."

[III. Author's View Argument ("*Solutio*")][53]

For these[54] and similar reasons we say that usury is unacceptable under any circumstances.

[IV. subarguments$_2$]:

[a. *Pro*]

[1] Whereas[55] the aforementioned chapter is not from the genuine Ambrose but from [ . . . Ambrose[56]], it [the argument] should be retracted.

[45] *Quod autem* . . .

[46] St. Ambrose, "De Tobia lib., c. xiv, 48; c. xv, 51" (Lefèvre 1902: 3 n. 4).

[47] *Item, in Deuteronomio* . . .

[48] *Item, lex Justiniana* . . .

[49] *Sed in contrarium* . . .

[50] *Item in Levitico* . . .

[51] *Item in Exodo* . . .

[52] *Et per novum Testamentum idem* . . .

[53] *Solutio*. Lefèvre's edition (1902: 7) prints this as "Solutio."

[54] I.e., the arguments *contra*, with which Robert agrees.

[55] *Unde* . . .

[56] The text has *sed Ambrosii cooperti*. The editor suggests that a missing piece of text may have claimed that the passage has been falsely labeled by another writer as Ambrose's work.

[2] And as for the authority of [the cited passage of] *Deuteronomy*,[57] it is to be explained . . .

[3] [Reply omitted.][58]

The next identified examples of the Latin recursive argument method,[59] chronologically, are to be found in the works of Alexander of Hales. Alexander (ca. 1185–1245) was an Englishman who went to Paris around the year 1200 to further his studies, and stayed there to become one of the most successful scholars and teachers of his day.

In the *Quaestiones Disputatae 'Antequem Esset Frater'* (ca. 1220–1236),[60] the earlier of his two major published works, Alexander uses the recursive argument method throughout. Like Avicenna—whose *Metaphysics* Alexander refers to indirectly in this work[61]—and like most other authors, Alexander does not explicitly number his subarguments fully (though he does frequently refer to the first or second Subargument$_1$ as '*primo*' or '*secundo*' in the Subarguments$_2$ section), but refers to them instead by repeating part of their content only, as in the examples of Avicenna's usage quoted in chapter 5.[62]

As fellow English students abroad in Paris, and then successful clerical scholars there, Robert of Curzon and Alexander of Hales, the verifiably

[57] *Et illa auctoritas Deuteronomii ita est exponenda . . .*

[58] No response is given to the Subargument$_2$ *Pro* that is based on the *Lex Justiniana*, though Robert presumably disagrees with it.

[59] I have restricted myself to those available in published editions. See also appendix B.

[60] Bieniak (2010: 143).

[61] Avicenna is cited explicitly in the *Summa Theologica* attributed to Alexander, e.g., Lib. I, Pars I, Inq. I, Tract. III, Quaest. I, Caput III 'De differentiis unius et unitatis' (Quaracchi 1924, 1: 117ff.), in which argument he also cites Aristotle and Algazel (al-Ghazālī), among others. It has long been believed that Alexander is responsible for writing or overseeing the writing of at least the first two volumes of the work, which was completed by others after his death, but Macy (2009) says that the *Summa* "is a compilation put together by his students after his death." In Example 6.5, I quote an argument from the very beginning of the first volume of the work.

[62] Alexander gives the Subarguments$_2$ in order of their appearance in the first list, but he is not careful about it; several Subarguments$_2$ sometimes are out of order, as shown in Example 6.4, and some are simply omitted—either because his discussion in the Author's View section makes further discussion unnecessary, or because discussion in those Subarguments$_2$ that he does give makes discussion unnecessary. Such irregularities are common in works of this genre that are based on live recordings of oral disputations (q.v. Little and Pelster 1934). Another reason for apparent irregularity is the practice of grouping the Subarguments$_2$ according to the shared main point that they make, which is found in many Latin recursive method works. Nevertheless the sequence is not of crucial importance for understanding the Subarguments$_2$ because Alexander refers to each Subargument$_1$ by citing the beginning of its text. Thomas Aquinas and Albertus Magnus religiously follow the proper order of the Subarguments$_1$ in their replies, which they typically number explicitly, though sometimes Subarguments$_1$ are replied to in groups in the Subarguments$_2$ section.

earliest Latin writers to use the recursive argument method in their own works, were both very prominent men and certainly knew each other—Robert resided in Paris from around 1195; Alexander was in Paris by about 1200, and was magister regens in the Faculty of Arts there in 1210. In 1213 Robert presided over a council there and in 1215 promulgated a Church decree regulating the University of Paris.[63] Alexander remained in Paris to the end of his life,[64] but Robert died on a crusade in 1218. Although it is unknown where they lived as students in Paris, we can speculate that they may have stayed, in turn, in the Collège des Dix-huit, the first college founded in Western Europe,[65] not only because it was the earliest but also because it was founded by an Englishman and may have been intended originally for English students studying in Paris, a pattern that was followed by donors from other places in later college foundations. If that was the intention of Jocius—though it is nowhere stated in the charter of his college[66]—it may explain why the earliest writers to use a form of the recursive argument method were Englishmen who were educated and taught in Paris at the end of the twelfth century and the beginning of the thirteenth.

### *Example 6.5*

**Alexander of Hales**

***Quaestiones Disputatae 'Antequem Esset Frater'***

Quaestio xlv, Disputatio II, Membrum I[67]

*De modo essendi Dei in rebus* 'On the Way God Exists in Things'

Part 1

[I. Main Argument (Question, Topic)]

[Whether God is omnipresent (*An Deus sit ubique*)][68]

[II. Subarguments$_1$]

[63] Lefèvre (1902: xv).

[64] Macy (2009).

[65] See chapter 3.

[66] See appendix C.

[67] Latin: Quaracchi 1960, 2: 760, 772–777.

[68] L. *membrum* 'part'; i.e., the first argument. The topic here has been supplied by the editors from the chapter-head summary.

[1] That he is omnipresent may be seen . . . (*Quod sit ubique, videtur*: . . . ) [full page of argument, including the author's proofs and citation of Anselm][69]
[2] Again (*Item*) [quotes Gregorius] . . .
[3] Again (*Item*) [quotes Ioannes Damascenus], . . .
[4] Again, on the same (*Item, ad idem*) [his own argument], . . .
[5] Again on the same (*Item ad idem*) [his own argument], . . .
[6] Again on the same (*Item ad idem*) [his own argument], . . .
[7] Again on the same (*Item ad idem*) [his own argument], . . .
[8] Against, that He is not omnipresent (*Contra, quod non sit ubique*): . . .
[9] Again [his own argument], . . .
[10] Further, it may be asked whether (*Praeterea quaeritur utrum*) [his own argument], . . .

[III. Author's View Argument]

I answer: I submit that God is everywhere. (*Respondeo: Concedo quod Deus est ubique.*)

[IV. $\text{Subarguments}_2$][70]

[8] To that which is objected . . . it is stated that . . . (*Ad hoc quod obicitur . . . dicendum quod . . .* )
[9] To that which is objected . . . it is stated that (*Ad hoc quod obicitur . . . dicendum quod . . .* ) . . .
[10] To that which is objected . . . I say that (*Ad hoc quod obicitur . . . dico quod . . .* ) . . .

## *Example 6.6*

## *Summa Theologica* attributed to Alexander of Hales

### Pars i, Inq. I, Tract. V, Sect. II, Quaest. I

### Chapter IV[71]

[I. Main Argument (Question, Topic)]

[69] The main presentation of the author's view in one of the $\text{Subarguments}_1$ with which he agrees is also found in Avicenna.

[70] As usual in Medieval Latin recursive method arguments, $\text{Subarguments}_1$ with which the author agrees are not replied to in the $\text{Subarguments}_2$.

[71] Latin text: Quaracchi 1924, 1: 269–270.

Whether God's foreknowledge imposes predestination on things[72]

[II. SUBARGUMENTS$_1$]

[1] On which affirmatively (L. *Ad quod sic*): . . .
[2] Again (L. *Item*) [quotes Anslem], . . .
[3] Again (L. *Item*), . . .
[4] Again (L. *Item*), . . .
[5] Again (L. *Item*), . . .
[6] Against (L. *Contra*), Augustine: . . .
[7] Again (L. *Item*), Anselm: . . .

[III. AUTHOR'S VIEW ARGUMENT]

I answer (L. *Respondeo*) . . . [Short paragraph; he agrees with the *contra* position.]

[IV. SUBARGUMENTS$_2$][73]

[1] Now, to the first objection (L. *Ad primum autem obiectum*), . . .
[2] In the same way the second is solved (L. *Eodem modo solvitur secundum*) . . .
[3] To the next one it is stated that (*Ad aliud dicendum quod*) . . .
[4] To the next one it is stated that (*Ad aliud dicendum quod*) . . .
[5] To the next one it is stated that (*Ad aliud dicendum quod*) . . .

Albertus Magnus (Albert the Great) is in many ways the most interesting of all the great scholars of the thirteenth century. He was a theologian-metaphysician like Alexander of Hales, and his *Summa Theologiae* is very similar to the *Summa Theologica* attributed to Alexander. This similarity extends to the numbering system Albert uses in most of the work. However, it is remarkable that in the beginning of the first volume (Tractatus I) Albertus uses a wide variety of numbering systems, ranging from explicitly numbering *all* of the SUBARGUMENTS in both lists,[74] to using numbers in only one of the two lists (as he does throughout his *Quaestiones super* 'De animalibus'; see EXAMPLES 1.1 and 6.7), to using no numbers at all, though as in Alexander's works the first SUBARGUMENT$_2$ is

[72] *Consequenter quaeritur utrum praescientia Dei imponat necessitatem rebus* 'Next it is asked whether God's foreknowledge imposes predestination on things.' The editors have evidently used this sentence to compose the chapter title given in small capitals in the published edition.

[73] As normal in Medieval Latin recursive method arguments, the author does not comment on SUBARGUMENTS$_1$ with which he agrees.

[74] Filhaut (1955: 25). Cf. Fakhr al-Dīn al-Rāzī, who also numbers both of the subargument lists. More study of numbering is needed.

typically numbered: *Ad primum* . . . 'As for the first . . . ' In arguments that have very long lists of subarguments, Albertus often groups them together in his replies (e.g., SUBARGUMENTS$_1$ numbers 4–7 are actually replied to with a single SUBARGUMENT$_2$), in which case he sometimes refers to the included subarguments by number even though no numbers are given in the actual SUBARGUMENT$_1$ list. The great variety of numbering styles in Albert's *Summa* is worthy of special study. It should be noted that the formal separation between SUBARGUMENTS$_1$ *Pro* and SUBARGUMENTS$_1$ *Contra* in the published edition is the creation of the editors. The original marks the first SUBARGUMENT$_1$ *Contra* solely as "*contra*" or "*sed contra*," as in other writers' works using the Latin recursive argument method. The practice of Albertus in the original text is followed here.

As mentioned above, Avicenna's *De anima*, one of the earliest and most influential scientific works translated from Arabic, uses the recursive argument method. Albertus seems to be the earliest European author to write new works on natural science using the recursive argument method. Not surprisingly, then, in his works on natural science topics, as also in his *Summa Theologiae*, Albertus refers constantly to the great Classical Arabic writers. He made original contributions to all of the sciences on which he worked, which include alchemy-chemistry, astrology-astronomy, botany, geology, metaphysics, theology, and zoology.

### *Example 6.7*

### ALBERTUS MAGNUS (ALBERT THE GREAT)

### *Quaestiones super* 'De animalibus'[75]

### BOOK 1

### QUESTION 4

### Whether Every Animal Breathes Air

[I. MAIN ARGUMENT]

"And the modes of animals," and so on.[76] We inquire here whether every animal breathes air.

[75] Latin: Filhaut (1955: 81–82).

[76] This is a reference to Aristotle's *History of Animals*, identified by Grant (1974: 682 n. 2) as I.1.487a.11–12. The *Quaestiones super* 'De animalibus' is a commentary on Aristotle's text. For other such text references, most of which are omitted here, see the notes to Grant's (1974: 682–683) full translation.

[II. SUBARGUMENTS$_1$]

[1] It seems that this is not so, because breathing is for the cooling of the heart and lungs; but air is warm and moist . . .

[2] Again, an animal drawing water lives from water. The sign of this is that if he is outside water, he immediately dies . . .

[3] The Philosopher [Aristotle] says the opposite.[77]

[III. AUTHOR'S VIEW ARGUMENT]

It must be said that there are certain animals that have a very warm heart, and nature has given them a lung which is like a fan . . . But the heart is an impassible member "for it is not susceptible to infirmity," according to the Philosopher in [Book] III of *On the Parts [of Animals]*.[78] But water and earth concern things materially . . .

[IV. SUBARGUMENTS$_2$] [Response] to the theories [*Pro*].[79]

[1] To the first, it should be said that although air is warm and moist, it is less intensely warm than the heart . . .

[2] To the second, it should be said that an animal which takes in water does not live from that water, because, according to Aristotle in the book *On Generation*, "we are nourished from the same things of which we are constituted." Therefore, since an animal . . .

In Latin Europe from the thirteenth century onward the recursive argument method "constituted the most commonly used format for the presentation of natural philosophy at universities. The great variety of questions on Aristotle's natural books was representative of what medieval natural philosophy was about."[80] In other words, for medieval science the recursive argument method was the literary "scientific method."

Finally, it would not do to omit an example from the *Summa Theologiae* by Thomas Aquinas, the greatest student of Albertus Magnus and

[77] Lat. *Oppositum dicit Philosophus* (Filhaut 1955: 82). Grant follows the modern Latin editors in not numbering the SUBARGUMENTS$_1$ with which Albert agrees. However, the original text does not number *any* of the SUBARGUMENTS$_1$ themselves; they have been numbered, in parentheses, by the editors, presumably to make clear their correspondence with the SUBARGUMENTS$_2$, which Albert does number in the text. The SUBARGUMENT$_1$ *contra* should thus be numbered "[3]" here even though it is not replied to by Albert, who agrees with Aristotle on the issue.

[78] The reference is to *On the Parts of Animals* 1.4.667a.33 (Grant 1974: 682 n. 4).

[79] L. *Ad rationes* (Filhaut 1955: 81–82). Grant (1974: 683) translates this as 'principal reasons', evidently referring to the omission of the argument with which Albert agrees, namely, the view of Aristotle. But, as previously noted, it is usual in Latin recursive method arguments to omit responses (SUBARGUMENTS$_2$) to SUBARGUMENTS$_1$ with which the author agrees.

[80] Grant (2007: 190).

the single most famous medieval scholastic. Although he did not write much on natural scientific topics, his impeccable attention to detail and logic makes his works models of scientific writing.

### *Example 6.8*

### Thomas Aquinas

### *Summa Theologiæ* 2a2æ. 64,5

#### Quaestio 64. De Homicidio 'On Homicide'

#### Article 5

Whether it is permitted for anyone to kill himself[81]

[I. Main Argument (Question, Topic)] Proceeding to the fifth [Article; i.e., Whether it is permitted for anyone to kill himself]

[II. Subarguments$_1$]

[1] It would seem that (*Videtur quod*) it is permitted for someone to kill himself. . . .

[2] Further (*Praeterea*), it is permitted for someone holding public power to licitly kill malefactors. But sometimes the one who holds public power is himself a malefactor. Therefore, it is licit for such a person to kill himself.

[3] Further (*Praeterea*), it is permitted for someone to voluntarily put oneself in lesser peril in order to avoid a greater peril; . . .

[4] Further (*Praeterea*), Sampson killed himself, as *Judges* records, but he is enumerated among the saints, . . .

[5] Further (*Praeterea*), *Maccabees*, Book II says that a certain Razis killed himself, choosing to die nobly rather than . . .

[6] But contrary (*Sed contra*) [to this] is what Augustine says, "We nevertheless understand that it is to man that the command 'Thou shalt not kill' refers . . . "[82]

[III. Author's View Argument] Reply (*Responsio*)

It is stated that (*Dicendum quod*) it is absolutely illicit to kill oneself, for three reasons. . . . [One paragraph]

[81] Latin: Lefébure (1975: 30–37), q.v. for a complete English translation.

[82] Because the author agrees with the Subargument$_1$ *contra* [6], he does not reply to it in the Subarguments$_2$, so there is no sixth Subargument$_2$.

[IV. SUBARGUMENTS$_2$]

[1] To the first (*Ad primum*), therefore, it is stated that homicide is a sin, not only because it is incompatible with justice; but also . . .

[2] To the second (*Ad secundum*) it is stated that one who holds public power can licitly kill wrongdoers, . . .

[3] To the third (*Ad tertium*) it is stated that man is constituted master of himself by [the fact of having] free will; . . .

[4] To the fourth (*Ad quartum*) it is stated that, as Augustine says, Sampson can only be excused for burying himself with his enemies . . .

[5] To the fifth (*Ad quintum*) it is stated that it is a mark of bravery that someone would not avoid being killed by someone in order to . . .

## The Oral and Literary Recursive Argument Methods

It is unknown whether or not the recursive argument method was first transmitted to medieval Europe via oral disputation and transferred to written form when the disputations were recorded, as argued by Makdisi,[83] or purely via the works of Avicenna (and possibly other Classical Arabic scholars) translated in the twelfth century. The tradition of recording and publishing oral disputations is known in both the Classical Arabic and the Medieval Latin traditions, as Makdisi shows at length, so it seems that *both* these means of transmission were involved.

Wippel defines "Disputed Questions"—recursive arguments—as those which have "the general scholastic format of question, opening arguments for one side and then one or more arguments for the other, a definitive response, and finally individual replies to the original opposing arguments."[84] According to Pelster, "The disputation as practised in the thirteenth century is a *discussion of a scientific question* between two or more disputants, of whom one undertakes the rôle of defender of a particular opinion, while the other or others raise objections and difficulties against this position."[85] While he argues that "the fully developed

[83] Makdisi (1981).

[84] Wippel (1985: 163).

[85] Similarly, Bazàn (1985: 42) summarizes the most salient parts of the recursive argument method (though he omits the *quaestio* itself, which is the MAIN ARGUMENT or "base case"), stressing the live, oral version: "La *disputatio*, avec ces trois notes d'exercice régulier, de discussion méthodique tripartite (arguments, solution, réponses aux arguments), et d'exercice comptant trois responsables principaux (le *magister* qui préside et détermine, l'*opponens* qui soulève des difficultés contre la thèse, le *respondens* qui clarifie préalablement le problème), est pratiquée

disputation can be traced back at least to the end of the twelfth century, he notes that "in the case of the questions about the middle of the twelfth century it cannot be definitely decided whether they are disputations in the strict sense."[86] That is a diplomatic way of saying that they are *not* examples of the recursive argument method. Fransen notes that it is only in the late twelfth or early thirteenth century that disputations in legal texts begin to include, *sometimes* ("parfois"), responses (SUBARGUMENTS$_2$) to the arguments (SUBARGUMENTS$_1$) proposed by the losing party,[87] that is, it was only then that the recursive argument method began to be used.

Pelster describes clearly the sharp difference between the earlier non-recursive argument form and the recursive argument method (though, following other scholars, he assumes that the latter developed organically in a straight line out of the former):

> The earliest form of the disputation, from which all the others sprang, would be the *quaestio in scolis*, as it is called in our MS f. 18r. The teacher put a question, a scientific problem: the student attempted to answer it: sometimes the position was reversed: the student asked the question and the teacher answered it.[88] A second stage is the disputation with a specially appointed *respondens*, who gave a preliminary solution of the question set, and then had to answer the objections raised by other participators. He was a different person from the teacher or master. The *opponens* or *quaerens* I was able to distinguish as early as about 1200 in the questions of Simon of Tournai, but not the *respondens*.[89]

---

systématiquement par les maîtres de la première moitié du XIIIe siècle." Note the closeness of this method to that of Vasubandhu, and to the Indian oral disputation method described by Saskya Paṇḍita, q.v. chapter 7.

[86] According to specialists who worked on the problem already a century ago, followed by Moore (1936: 26–27) and many others, the first appearance of the recursive argument method in Latin literary works dates to ca. AD 1200. "But students disagree on the time at which the *disputatio* became separated from the *lectio* to form a distinct scholastic exercise," and some "believe that the *disputatio* did not become a distinct exercise until very late in the twelfth century or early in the thirteenth century" (Moore 1936: 46–47). This shows that already long ago there was considerable unclarity about the scholastic method; it has since grown tremendously.

[87] Fransen (1985: 244–245).

[88] Pelster (1934: 31) describes the typical twelfth-century type of *disputatio*, such as that seen in the *Disputationes* of Simon of Tournai in the late twelfth century, which has no SUBARGUMENTS$_2$ section and was thus *not* the recursive argument method, as he remarks.

[89] Pelster (1934: 31–32), who then gives several early thirteenth-century examples, including from Alexander of Hales's question in *De miraculo*, where he says, "Concedebat hoc respondens et dicebat," clearly pointing to the presence of the *respondens*.

It is only in the thirteenth century, says Pelster, that "the practice [of including a respondens] becomes general."[90]

Grant[91] stresses the vital importance of the recursive argument method for the development of a full scientific culture in Western Europe. Similarly, Bazàn says, "the *disputatio* is the very heart of medieval scholasticism, and developed as the scientific spirit of the Middle Ages matured. The *disputatio*, as *scientific method*, is established as the result of the mastery of logic and the development of a speculative spirit among medieval thinkers. . . . The *disputatio* is thus an expression and product of the self-awareness of medieval scientific culture."[92]

Among typical differences between a *reportatio* of an oral disputation and a "Master's definitive written version," Wippel notes "a sense of being unfinished, a certain lack of tidiness in mere *reportationes*. For example, one may find replies to objections without the objections themselves appearing where they should, or perhaps, even without their being present in the Quodlibetal question at all as it now survives."[93] This characteristic feature of the *quaestiones disputatae* of Alexander of Hales indicates that they had a similar oral origin.[94]

[90] Little and Pelster (1934: 29; cf. 31–32).

[91] Grant (2007: 188).

[92] Bazàn (1985: 21), my translation. He adds, "Ainsi, dans le traité *De fallaciis* attribué à saint Thomas, elle est définie comme 'actus syllogisticus unius ad alterum ad aliquod propositum ostendendum'. Cette définition est adéquate pour caractériser la *disputatio* en tant que *méthode scientifique*. Elle la situe clairement dans le genre des actes scientifiques, signale l'instrument dont on s'y sert, le syllogisme, indique le caractère dialogique de l'activité, qui implique la présence d'au moins deux personnages (*opponens* et *respondens*), et met en évidence le but de cette recherche, à savoir la démonstration (*ostensio*) de la vérité touchant une question soulevée ou proposée (*propositum*)." Bazàn (1985: 22–23), emphasis in the original.

[93] Wippel (1985: 190).

[94] See the discussion above in this chapter.

## — Chapter Seven —

# India, Tibet, China, Byzantium, and Other Control Cases

The first civilization in the world to develop a full scientific culture was medieval Western Europe. It led directly to the scientific revolution—during which some changes to the details of the constituent elements took place—and continued on down to modern science. But the reason such a "scientific culture complex" developed uniquely in Western Europe has remained unclear.

The essential elements of medieval science were introduced to Western Europe via Classical Arabic civilization. Previously, the same elements had similarly been introduced to the Islamic world.[1] There the Classical Arabic form of science developed and was eventually transmitted, together with other cultural elements, to Medieval Latin culture, mainly via literary translations done in newly re-Christianized Spain. In Western Europe these foreign cultural elements, despite their various origins, were eagerly accepted, blended with native institutions, and incorporated into the rapidly changing intellectual culture of the time.

What, precisely, were these essential elements that produced, uniquely, a full scientific culture in Western Europe? In order to answer that question it is necessary to compare the constituent elements in the one culture in which it did develop with other cultures that had the same constitutive elements but did not develop science. These are the control cases.

It is now well established that the identifiable predecessor of the early modern form of science that developed during the scientific revolution goes back to Medieval Latin science, which in turn developed out of Classical Arabic science. In the latter case, despite possession for centuries of all the constituent elements of a full scientific culture, such a culture did not develop there, and eventually science largely withered. In

[1] There is nothing specifically Arab or Islamic about Classical Arabic science. It is called "Classical Arabic" science in this book because some of it was done by non-Muslims, natural philosophy was early condemned as "un-Islamic," and almost all of it was written in Classical Arabic—a rich, complex, flexible language well suited to natural philosophy.

Medieval Latin civilization, by contrast, a full scientific culture developed within a few decades of the introduction of the very same elements that were at the time still present in Classical Arabic civilization. The different outcome in these two cases is both striking and puzzling.

There are, however, further examples. Long before the college and the recursive argument method appeared in the Islamic world and Western Europe, they were known in India, Tibet, and China. Nevertheless, science (that is, beyond generic human "science") never developed in those cultures, not to speak of a full scientific culture. This might lead some to the hasty conclusion that Eurocentric scholars' view is right: the only thing that matters is Aristotle and other European things. However, such a conclusion simply cannot be supported on the basis of the evidence. Although it seems that the enthusiastic reception of the elements of a full scientific culture in Western Europe and the successful establishment of that culture complex could conceivably be due partly to a particular European cultural characteristic, such as an unusually acute curiosity about the world, it has been argued that most Asian cultures at one time or another had individuals who were evidently as curious about the world as any European. In addition, the fact of the transmission of the entire complex of ideas at about the same time to approximately the same part of Europe (northwestern France and southeastern England) could theoretically also have been decisive—and in any case cannot be overlooked—but the fact remains that most of the same elements entered the Islamic world at about the same time too, so "simultaneity" fails as an explanation for the different outcomes in the two civilizations. This chapter constitutes an attempt to analyze why a full scientific culture developed only in Western Europe and only after the mid-twelfth century, and did not develop anytime or anyplace else.

## Graeco-Roman Antiquity

The ancient Greeks, especially Aristotle and his followers, developed the basic idea of science. Aristotle applied it to nearly everything he could think of,[2] and in so doing also divided knowledge up into separate fields. After his death, Aristotelianism continued on as one among several schools of thought in the ancient Mediterranean world. The first culture to

[2] Aristotle largely restricted the use of mathematics to the "mathematical sciences." There were a number of reasons for this, but one of its consequences seems to be that the rest of the sciences were not subjected by him to quite the same logical rigor that the "mathematical sciences" were.

develop a scientific tradition per se was thus the Graeco-Roman world. If possession of the works of Aristotle and the other ancient Greek thinkers alone were sufficient to produce a full scientific culture, it should thus have happened in classical antiquity between the lifetime of Aristotle (384–322 BC) and the edict of toleration of Christianity in the Roman Empire in AD 313, a period of more than six centuries.

But it did not. No doubt, when the Graeco-Roman world converted to Christianity, the ancient philosophical books, including some of the scientific parts of them, were considered suspect. Yet there was no great book burning, and some scholars did continue to read and study the ancients for many centuries more in Greek-speaking areas of the Mediterranean region. So the failure of the Graeco-Roman world to develop a full scientific culture would seem to be rather puzzling. However, there appear to be some fairly good reasons for this particular failure.

In antiquity nearly all education was private, conveyed to children by hired tutors. There were some grammar schools, but there was no education *establishment* of any kind. There were a few highly localized advanced academic institutions, such as the fabled Library of Alexandria and the Academy at Athens, but the former disappeared centuries before the Arab conquest, and the latter, supposedly closed by Justinian in 529, had actually ceased to exist centuries earlier.[3] In any case, there was no widespread system of higher education to ensure that knowledge was passed on from generation to generation within Graeco-Roman civilization. Under such conditions the fact that any of the ancient Greek writers' works have been preserved at all is practically miraculous, and indeed, for many of them we have nothing but a few brief quotations preserved by chance, typically in a rather chatty or "light" work, or in something of interest to the early Christians. It is surely not difficult to see, then, the importance of having a kind of institution that is essentially permanent and devoted exclusively to research, to teaching, and to the preservation and dissemination of the accumulated wisdom of one's predecessors in the form of books and other publications. For the natural sciences, this is even more important than for other fields because of the great amount of precise detail that is necessary in order for science to progress.

We must also consider what "doing science" meant in practice. In Aristotle's mind, science meant analysis through reason, logic, and mathematics as well as careful, exhaustive argumentation: "We must, with a view to the science which we are seeking, first recount the subjects that

[3] Szlezák (2002: 45).

should be first discussed. These include both the other opinions that some have held on certain points, and any points besides these that happen to have been overlooked. For those who wish to get clear of difficulties it is advantageous to state the difficulties well; for the subsequent free play of thought implies the solution of the previous difficulties, and it is not possible to untie a knot which one does not know."[4]

The amount of work he put into becoming an expert debater is, by his own account, astounding: "We should select also from the written handbooks of argument, and should draw up sketch-lists of them upon each kind of subject, putting them down under separate headings, e.g. 'On Good', or 'On Life'—and that 'On Good' should deal with every form of good, beginning with the essence. In the margin, too, one should indicate also the opinions of individual thinkers, e.g. that Empedocles said that the elements of bodies were four; for any one might assent to the saying of some reputable authority."[5]

Nevertheless, the Greeks had no overt, rigorously systematic method of formal analysis.[6] Aristotle's scientific works are written according to the treatise format, as are the works of most of the other ancient writers on scientific topics, though some—most famously Plato—preferred the dialogue format. Aristotle discusses oral debate in great detail, especially in the *Topics*, but he nowhere describes anything like the recursive argument method. He does suggest something that sounds similar, as a way of preparing for a debate: "Before maintaining either a thesis or a definition the answerer should try his hand at attacking it himself; for clearly his business is to oppose those positions from which questioners demolish what he has laid down."[7]

However, nowhere in any ancient or medieval Greek work, on any topic, or any early medieval Latin work, is anything like the recursive argument method used. Nor is there any other literary form used especially for works of "natural philosophy." As far as the record of ancient science is concerned, therefore, there was no "scientific method."

[4] *Metaphysics*, iii.1 (Barnes 1984, 2: 1572).

[5] *Topics*, i.14 (Barnes 1984, 1: 175).

[6] Of course, Aristotle's works may be characterized as being obsessed with logical analysis (often to the exclusion of practically anything else); that is what gives them their "scientific" character, whatever their factual content. But content, including logic, must be distinguished from formal argument structure, as shown in chapter 2.

[7] *Topics*, viii.9 (Barnes 1984, 1: 270). One of the most remarkable points about this description is its explicit reference to written handbooks on debate, written analysis, and written notes or glosses on texts. This unambiguously points to Greek intellectual culture's focus on the written word no later than Aristotle, who died in 322 BC.

The ancients had Aristotle's works, which contain the idea of science and the scientist's self-consciousness, as well as the distinction between science and non-science, and they certainly did develop a sophisticated form of nascent science. But perhaps the works of Aristotle and his successors were too difficult to understand or too dry. One can hardly blame ancient Graeco-Roman intellectuals for preferring other thinkers and writers, especially Plato, who has an engaging literary style and a metaphysical philosophy that was much easier to adapt to Christianity when that religion became the dominant thought system of the region.

However, such possible faults of Aristotle were there from the beginning and were still there when his works were involved in the development of science in Classical Arabic civilization and again in Medieval Latin civilization. Therefore, they cannot be the reason for the failure of a full scientific culture to develop in the ancient Graeco-Roman world. The logical conclusion is that Aristotle *alone*, even along with all the rest of ancient Greek literature, is insufficient for the development of a full scientific culture.

## India

The *vihāra* 'college' was born in Central Asia during the Kushan Empire period (ca. 50 BC to ca. AD 225), and spread all around the empire, including throughout northern India, where it soon replaced the earlier native Indian *saṅghārāma*.[8] In some cases, the *vihāra* or *mahāvihāra* 'great *vihāra*' developed into an extensive university-like complex. The largest and most famous examples were three medieval *mahāvihāras* that developed in Eastern India, in the area of what is now Bihar and Bengal: Nālandā, which is partly excavated, located not far from the ancient city of Pāṭaliputra, modern Patna; Vikramaśilā (or Vikramaśīlā), and

[8] The *saṅghārāma* grew organically within early Buddhism as it developed in India. Its earliest precursors were early shelters, or *ārāmas*, under which the monks took refuge from the annual monsoon. Over time these became increasingly permanent *saṅghārāmas*, either buildings newly erected for the purpose or modified existing structures that were donated by wealthy Buddhist laymen and laywomen. After the introduction and rapid spread of the college under Kushan patronage, the term *saṅghārāma* eventually came to be synonymous with the term *vihāra* (despite the actual differences), and finally the new *vihāra*, including its particular form, function, and endowment, as well as its name, supplanted the older *saṅghārāma*. The old term *saṅghārāma*, however, was affectionately retained by some medieval Buddhist writers, most notably Hsüan Tsang, who applied it to famous institutions even in Central Asia, where they were probably known as *vihāra*s from the very start. Cf. Dutt (1962).

Odantapurī (or Odantapura).[9] In addition, the greatest center of learning in "Indian" Gandhāra during classical antiquity was the city of Taxila, or Takṣaśilā (located near what is now Islamabad in Pakistan), where the ruins of several important early Buddhist *vihāras* have been excavated. There teachers and students of Buddhism and other major Indian traditions were to be found.[10] Having adopted the Central Asian college (*vihāra*), the Indian Buddhists thus had large, permanent institutions devoted to higher education.[11]

The Indians also had a version of the recursive argument method, which was likewise introduced from Buddhist Central Asia during the period of Kushan Empire rule over northern India.[12] In addition, Indians had some of the other essential elements of a potential full scientific culture, including advanced mathematics and astronomy—much of which was indebted to Greek astronomy.

However, the Indians did not have the crucial *idea* of science, a scientific approach to the world, and the idea of scientific fields, which they would have gotten from Aristotle if they had followed up on their intensive, long-term contact with the Greek world from at least the campaign of Alexander onward,[13] or if in the Middle Ages they had read the Arabic translations of Aristotle eventually known to Indian Muslims. In fact, scientists of the Classical Arabic and Medieval Latin cultures, such as al-Bīrūnī and Roger Bacon, are explicitly conscious, and rather vehement, about the difference between scientific work and nonscientific work, regardless of the actual topic being treated. One sign of true science is that it is not mixed up with pseudo science.[14] Al-Bīrūnī emphatically points out this distinction in his book *India*. Although it is mainly a descriptive ethnography of Indian religious ideas, it was in his mind nevertheless a scientific work. By contrast, he says,

[9] Dutt (1962: 328ff.).

[10] Dutt (1962: 211–221).

[11] Several of them survived, though in reduced circumstances (intellectually as well as physically), down to their destruction by Afghan Muslims in the Ghaznevid period. It may be argued that the adoption of Hindu Tantrism and other beliefs and practices by some followers of Mahayana Buddhism made it not much different from popular forms of Hinduism; the resulting spread of Hindu ideas at the expense of genuine Buddhist ones would thus be to some extent responsible for the colleges' intellectual decline (Dutt 1962). However, this scenario does not explain the spread and flourishing of the very same Tantric form of Mahayana Buddhism in Tibet, Mongolia, and China, with many thousands of *vihāras* down to the twentieth century. This particular aspect of the problem thus remains open.

[12] See the example by Kamalaśīla in the following section.

[13] Cf. chapters 3 and 4.

[14] Some medieval scientists abhorred astrology, some alchemy, some both, and some neither.

> The Hindus had no men . . . both capable and willing to bring sciences to a classical perfection. Therefore you mostly find that even the so-called scientific theorems of the Hindus are in a state of utter confusion, devoid of any logical order, and in the last instance always mixed up with the silly notions of the crowd, *e.g.* immense numbers, enormous spaces of time, and all kinds of religious dogmas, which the vulgar belief does not admit of being called into question. Therefore it is a prevailing practice among the Hindus *jurare in verba magistri*; and I can only compare their mathematical and astronomical literature, as far as I know it, to a mixture of pearl shells and sour dates, or of pearls and dung, or of costly crystals and common pebbles. Both kinds of things are equal in their eyes, since they cannot raise themselves to the methods of a strictly scientific deduction.[15]

Despite the many important scientific achievements of Indian scholars over the ages, and the important contributions their mathematics and astronomy made to the development of science in Classical Arabic civilization and medieval Western Europe, a full scientific culture never appeared in India.

## Tibet

The college, in the form of the Buddhist *vihāra*, was transmitted to Tibet when Buddhism was introduced there in the seventh century. The first historical *vihāras*, including at least one in the capital, Rasa (now Lhasa), were built during the reign of Emperor Khri Sroṅ Rtsan (d. 649/650).[16] The first large *vihāra* in which teaching, research, and translation activities were actively pursued was Samye, which was dedicated in ca. AD 764.[17] In addition to many Buddhist texts, which were mostly translated from Sanskrit and secondarily from Chinese, some Greek medical thought was transmitted, apparently via Bactria.[18]

The recursive argument method was introduced to Tibet during the imperial period by the Indian Buddhist scholars Śāntarakṣita (ca. 725–788) and his pupil Kamalaśīla (ca. 740–795). In Kamalaśīla's

[15] Al-Bīrūnī (1910, 1: 25).

[16] Beckwith (1993). He is popularly known as "Srong-btsan Sgampo."

[17] Walter and Beckwith (2010); cf. Walter (2009) and Beckwith (1984b) on the plan of Samye and its significance.

[18] Beckwith (1979), Yoeli-Tlalim (2010).

*Madhyamakāloka* a brief mention of the topic is followed by two lists of arguments,[19] of the type referred to in treatises about logic and debate in Sanskrit as *pūrvapakṣa* (in Tibetan, *phyogs sṅama*) and *uttrarapakṣa* (in Tibetan, *phyogs phyima*).

The Indian recursive argument method is simpler than the developed Central Asian method, in that each of the two recursive lists always represents only one view. The lists also can be very long. In Kamalaśīla's method, the entire work is divided up into a *pūrvapakṣa* section consisting of different subarguments supporting one main argument about a common topic by the opponent, and an *uttrarapakṣa* section consisting of the arguments of the author replying in order to the opponent's points.

The very beginning of Kamalaśīla's argument in his *Madhyamakāloka* is analyzed below (see EXAMPLE 7.1), largely following the treatment by Keira.

## *Example 7.1*

### KAMALAŚĪLA

### *Madhyamakāloka*[20]

[I. MAIN ARGUMENT (QUESTION, TOPIC)]

If one considers whether it can be proved, by means of Scripture or reason, that all constituents[21] have no intrinsic nature . . . [22]

[II. SUBARGUMENTS$_1$][23]

[19] For an example of this argument form, see the study and partial translation of the *Madhyamakāloka* by Ryusei Keira (2004). However, some parts of the argument are unfortunately omitted in Keira's translation, as noted below.

[20] Tibetan text: Derge *Dbuma, sa* (Delhi 1982–1985, 107: 266ff.), folios 133a.7–171a.2; the text quoted here is on folios 134b.7–171a.2 (Keira 2004: 223–230); for a more or less full translation, see Keira (2004: 89–117).

[21] Sanskrit *dharma*.

[22] Keira (2004: 89) skips the MAIN ARGUMENT (TOPIC OR QUESTION), which occurs on folio 133a.7 of the Derge edition, and instead gives the following topic of the argument as a chapter heading: "The Establishment of the Absence of Intrinsic Nature in Reality by Means of Direct Perception." This does not correspond to the Tibetan and seems to have been invented by Keira to describe the argument's contents.

[23] These arguments are all *contra* the author's view; they are regarded as Dharmakīrti's positions (Keira 2004: 90 n. 133). As a heading for them, Keira (2004: 89) has "OBJECTIONS (*pūrvapakṣa*)" in the translation, and "*Phyogs snga ma* (*pūrvapakṣa*)" in the transliterated Tibetan text section (Keira 2004: 223). However, neither corresponds to anything in the Tibetan text.

[1] It also[24] cannot be [proved] by means of reason. Thus, one would not understand by means of direct perception that all entities without exception are empty . . .

[2] If that direct perception were also to have an intrinsic nature . . .

[3] Furthermore, if one holds a place to be free of another object . . .

[4] Also, [the assumption] that this [absence of intrinsic nature] is perfectly cognized too by the yogic perception of the Omniscient one . . .

[III. SUBARGUMENTS$_2$][25]

[1] Someone having said,[26] "It also cannot be [proved] by means of reason. Thus, one would not understand by means of direct perception that all entities without exception are empty . . . ," we reply, . . . [27]

[2] Someone having said, "If that direct perception were also to have an intrinsic nature," and so on, . . .

[3] Someone having said, "If one holds a place to be free of another object" . . .

[4] Someone having said, "With respect to that, yogins have understood it perfectly through direct perception" . . .

In this version of the recursive argument method, "Kamalaśīla considers all possible arguments that could be or have been brought forward against the assumption that all things lack an intrinsic nature. Besides the proof by Scripture (*luṅ*, *āgama*), he considers the proof by reason (*rigspa*, *yukti*). Under the heading of *yukti* Kamalaśīla discusses all other possible arguments such as perception and inference. These possible *pūrvapakṣas* [or SUBARGUMENTS$_1$] are then refuted in the *uttarapakṣa* [or SUBARGUMENT$_2$] section, one by one and more or less in the same sequence."

[24] Keira does not note that he has skipped the preceding section of SUBARGUMENTS$_1$ based on scripture ("To begin with, it cannot [be proved] by means of scripture"), and the SUBARGUMENTS$_2$ corresponding to them (which follow the section of SUBARGUMENTS$_1$ based on reason). In order to avoid terminal confusion I have, regrettably, therefore also omitted the SUBARGUMENTS$_1$ based on scripture and the SUBARGUMENTS$_2$ corresponding to them.

[25] The SUBARGUMENTS$_2$ all represent the view of the author. Keira (2004: 93) gives here "REPLIES (*uttarapakṣa*)" in the translation, and "*Phyogs phyi ma* (*uttarapakṣa*)" in the transliterated Tibetan text section (Keira 2004: 224), but again, neither corresponds to anything actually written in the text.

[26] The Tibetan construction marking the beginning of all four SUBARGUMENTS$_2$ is the same: *gang yang . . . zhes*, literally, 'someone ["] . . . ["] saying-unquote'. The similarity to the typical usage in Classical Arabic arguments is striking.

[27] The Tibetan in quotes is an exact copy of the SUBARGUMENT$_1$ phraseology. The remaining SUBARGUMENTS$_2$ are very close but not identical to their corresponding SUBARGUMENTS$_1$.

The entire long argument can ultimately be divided up into the following subsections (each of which includes numerous arguments), according to their content:[28]

I. SUBARGUMENTS$_1$ (Skt. *pūrvapakṣa*, arguments *Contra*)

[1] according to Scripture (Tib. *luṅ*, Skt. *āgama*)

[1.1] Scripture quotations

[2] according to reasoning (Tib. *rigspa*, Skt. *yukti*)

[2.1] according to perception
[2.2] according to inference

II. SUBARGUMENTS$_2$ (*uttarapakṣa*, arguments *Contra* the *pūrvapakṣa*)

[1] according to Scripture

[1.1] Scripture quotations

[2] according to reasoning

[2.1] according to perception
[2.2] according to inference

One of the distinctive features of this method is that there is no discrete AUTHOR'S VIEW ARGUMENT section: the SUBARGUMENTS$_2$ (*uttarapakṣa*) represent Kamalaśīla's view. It is therefore very much like the method in the *Aṣṭagrantha*,[29] the earliest preserved Central Asian Buddhist text that uses a form of the recursive argument method. Kamalaśīla's method, the normal method of Indian Buddhist scholastic texts, thus appears to have been introduced from Central Asia under the Kushans along with the *vihāra*, as noted above.[30]

This Indian argument structure is described by Saskya Paṇḍita in his *Mkhaspa la 'jugpa'i sgo*: "The non-Buddhist sectarians say that after the proponent recites in a single continuous statement a [long] series of different *pūrvapakṣas* such as one hundred or two hundred of them, the respondent [should] repeat that [long series of] *pūrvapakṣas* without mixing up their order. Then [the respondent] should refute [that]

[28] Helmut Krasser (personal communication, December 2009); I have added the material in square brackets and have reformatted everything to fit the style of this book. All errors are of course my responsibility.

[29] See chapter 4.

[30] See chapter 3.

ordered series by means of [his] *uttarapakṣas*. But if [the respondent] is unable to state such [a long series] of *pūrvapakṣas*, there will occur the defeat situation called 'not repeating.'"[31]

Saskya Paṇḍita further clarifies that this kind of argumentation is actually between two positions, or it should be:

> The debating back and forth by the proponent and respondent [is to be done] for the purpose of upholding their own doctrine. . . . But to debate heedlessly with no wish to uphold a doctrine and with the aim of testing each other's knowledge, like the argument of fools, is not the debating of noble persons, but it is merely crafty people amazing the ignorant. . . . If [a debater] does not take sides with any philosophical position, by determining what his initial position (*pūrvapakṣa*) [is], [his stand] will usually be [seen to be] the reasoning of an ignorant person.[32]

The absence of an AUTHOR'S VIEW ARGUMENT section is not the only significant difference between the Indian method and the classic recursive argument method known from later Central Asian Buddhist works, Classical Arabic works, and Medieval Latin works. Kamalaśīla's *Madhyamakāloka* as a whole is structured as a single large recursive argument. This is unlike the classic recursive method, in which problems are first broken down into the smallest discrete parts, or MAIN ARGUMENTS, which are typically called 'questions'. Each question, which is usually explicitly mentioned as such, is then analyzed in its own recursive argument. The Indian method—as shown by Kamalaśīla's *Madhyamakāloka* and other examples, and described in the quotes from Saskya Paṇḍita above—thus has a rather different character from the classic method, which developed later in Central Asia after the simpler early version of the method had been transmitted to India.[33]

The argument structure used by Śāntarakṣita and Kamalaśīla in their texts written in Tibet is seen earlier in Indian scholastic works of the Mādhyamika tradition, such as the *Vigrahavyāvartanī*, a little work

[31] Jackson (1987: 359), who notes this is "the sixteenth occasion of defeat mentioned in the *Nyāya Sūtras* and the fourteenth in Sa[skya]-paṇ[ḍita]'s list" (Jackson 1987: 447 n. 201). Saskya Paṇḍita's text continues with variations discussed by Dharmakīrtī in his *Vādanyāya* (Jackson 1987: 447 n. 202). For a detailed analysis of a Tibetan scholastic debate, see Jackson (1987: 196–203).

[32] Jackson (1987: 340–341).

[33] See chapter 4.

attributed to Nāgārjuna.[34] The arguments in these texts are organized according to the same Indian recursive argument method.

In addition to what thus became the traditional Indian method, a later Central Asian form of the recursive argument method, essentially the classic Central Asian type,[35] was introduced to Tibet at the same time, the late eighth century, via translation into Tibetan of Vasubandhu's *Abhidharmakośabhāṣya*.[36] However, neither it nor the typical Indian method was followed by later Tibetans.

The Tibetans adopted yet another method instead. The traditional Tibetan scholastic argument method, usually referred to as *dgag gźag spaṅ gsum*—the three [-part argument consisting of] *dgag* 'objections' [of others, and arguments against them], *gźag* 'establishing' [one's own view, and] *spaṅ* 'refutation' [of additional objections to one's view]—was apparently introduced by the influential teacher Rṅog Lotsāva Blo-ldan Śes-rabs (1059–1109), who seems to have developed it himself.[37] This structurally quite different argument type, which is *not* a form of the recursive argument method,[38] was adopted by one of the most famous scholars in Tibetan history, Saskya Paṇḍita (1182–1251),[39] who uses it in several of his most important works. It thus became a mainstream tradition in Tibet.[40] It was widely used by later scholars and remains well known due to its use as the main argument structure in the textbooks of the dominant modern sect of Tibetan Buddhism, the Dgelugspa.

[34] Bhattacharya et al. (1978). It would thus be traditionally dated to about the second century AD, in the Kushan Empire period.

[35] See chapter 4.

[36] Translated by Jinamitra and Kaba Dpal-brtsegs Rakṣita, both of whom flourished in the second half of the eighth century (Hermann-Pfandt 2008: 383). For examples from this text, see chapter 4.

[37] Rṅog Lotsāva had studied in Kashmir and might possibly have learned it there, but this Tibetan argument method is not attested in Indian works, so it is probable that he (or another early Tibetan scholar) innovated it. He is also said to have written a summary of Kamalaśīla's *Madhyamakāloka*; see Keira (2004: 17 n. 41).

[38] It is obviously a tripartite argument, for which reason I previously argued (Beckwith 1990) that it is directly comparable to the Arabic and Latin scholastic methods. However, among other differences from them it appears that the parts of the Tibetan argument do not contain recursive lists. Since the key feature of the recursive argument method is its *recursive* arguments, not the number of parts in the argument, the Tibetan method is not a version of the recursive argument method. My earlier view must accordingly be rejected.

[39] Saskya Paṇḍita is also known for having arranged Tibet's peaceful submission to the Mongols under the Great Khan Ögedei in 1247 (Beckwith 2009: 191).

[40] According to Dreyfus (2008), and *pace* Stein (1972: 160–163), it bears no relation to the oral form of debate used by Tibetan monks today (typically also called *dgag gźag spaṅ gsum*), which he says actually has the repeated Question : Answer (ABABAB, etc.) type of dialogue structure widely found in ancient and medieval literatures, as noted above. However, the analysis of Tibetan debate by Jackson (1987: 196–203) suggests that formal Tibetan debate *does* follow the ancient Indian model. Therefore, this question would appear to require further research.

In short, the Tibetans acquired some Greek science from the Central Asian Buddhists, and from Buddhist texts they got the Central Asian and Indian forms of the recursive argument method. They also adopted the Buddhist college (*vihāra*). Like the Indians, the Tibetans thus had available to them several of the essential elements of a full scientific culture, and some were actually adopted in Tibet. It is true that because Tibetan medicine adopted elements of Greek (as well as Indian and Chinese) medical science, it was in some respects more advanced than that of its neighbors to the north and east, and it has traditionally been highly esteemed by the Chinese. But the Tibetans did not have or get the *idea* of science, and although they were exposed to the developed recursive argument method in the *Abhidharmakośabhāṣya*, they seem never to have used the recursive argument method themselves, but instead developed their own argument method for use in academic works. Tibet produced very little science or scientific technology,[41] and never developed a full scientific culture.

## China

The Buddhist college (*vihāra*) was transmitted to China very early. It may have existed already in Kansu between the mid-first century BC and the mid-first century AD, but in any case sober historical sources solidly attest its existence in China proper by the second half of the first century AD.[42] The Chinese occupied much of Central Asia itself militarily for a lengthy period in antiquity and again in the early Middle Ages, and thus bordered on the Graeco-Bactrian culture zone. Yet they did not acquire any texts of Aristotle or other works of Greek science at that time. Even a half millennium later, in the Mongol period, despite the presence in China of many educated Muslim scholars and scientists, who certainly knew much Graeco-Arabic science and transmitted some technical knowledge from it, the works of Aristotle were not translated into Chinese. Other than to improve some native Chinese scientific technology, mainly in astronomy-astrology and medicine, therefore, Graeco-Arabic science had no effect on Chinese science before the appearance of European ships on the China coast in the sixteenth century.

The developed recursive argument method appears in Chinese in the early Central Asian *Vibhāṣa* attributed to Sitapāṇi (or Śītapāṇi), which

[41] On what they did develop, see White (1960), and cf. the remarks on advanced Tibetan automata and other metallurgical technology in Beckwith (1993: 185).

[42] Zürcher (2007: 26ff.).

was translated in AD 383 by Saṅghabhadra;[43] in the Kashmiri *Mahā-vibhāṣa* translated in AD 437–439 by Buddhavarman, and again, from another recension, by Hsüan Tsang in AD 656–659; and also in Vasubandhu's *Abhidharmakośabhāṣya*, a famous, very important text that was translated,[44] studied, and read by many in China. But Chinese Buddhist authors rarely, if ever, used the recursive argument method in their own personally authored Chinese works.[45] Moreover, the college in China was used largely for religious purposes and never developed into a genuinely *academic* institution as it did in Buddhist Central Asia and India, and later in the Islamic world. The Chinese did produce some science, and made many technological discoveries that were historically very important,[46] but they never really got the idea of science, and certainly never developed a full scientific culture complex. The same is true of the Japanese, Koreans, and Vietnamese, the most heavily Sinified of the neighboring peoples.

## Early Medieval Western Europe

Educated ancient Romans normally learned Greek and read Greek works in the original, so they did not translate many of them into Latin. In late antiquity, however, several important Greek works, including two important logical works by Aristotle, the *Predicamenta* 'Categories' and *De interpretatione* 'On Interpretation', and Porphyry's *Isagoge*, an introduction to Aristotelian logic, were translated into Latin by Boethius and were continuously known throughout the Middle Ages. The idea of science suffuses all of Aristotle's works, so it should thus have been transmitted to post-Roman Western Europe. Nevertheless, Latin Europe did not develop more than extremely marginal science, not to speak of a full scientific culture, at that time.[47]

[43] Cox (1998: 232).

[44] Once in AD 565 by Paramārtha and once in 653 by Hsüan Tsang; two Chinese manuscripts of the text have also been found in Tun-huang; see Cox (1998: 272) for further references. See also chapter 4.

[45] I have not been able to find any reference to examples of such use. I have been informed that the method is used in a sixth-century Chinese commentary, but I have not been able to obtain a legible copy of the study in question (Plassen 2002); cf. Plassen (2007: 120–122, 121 n. 16). One of the difficulties faced in attempting to analyze one of the arguments in these works is their use of explicit numbers as references to sections, subsections, sub-subsections, etc., without making clear the structure of the text per se. The same is true of Tibetan scholastic texts. Although both traditions have developed analytical tools to help readers follow the arguments, they are still daunting to uninitiated readers like me.

[46] The classic study is Needham (1954–2008).

[47] See Eastwood (2002), McCluskey (1998), Lindberg (2007: 194ff.). Cf. chapter 1, note 17.

In the early to mid-twelfth century many of Aristotle's most important works became available in excellent Latin translations made by James of Venice and others directly from the Greek. However, still nothing happened; Aristotle and other ancient Greek thinkers are exceedingly rarely cited by Medieval Latin writers until well into the thirteenth century. Despite extensive knowledge of Aristotle, a full scientific culture complex still did not develop in Western Europe until *after* the transmission of the recursive scholastic method and the college in the second half of the twelfth century, the same period when Avicenna's Aristotelian works were translated from Arabic, and especially in the late twelfth and early thirteenth centuries, when Aristotle's own works were translated from Arabic and Greek along with the great Classical Arabic commentaries on them.[48]

## The Byzantine Empire

The conversion of the Eastern Roman Empire to Christianity in late antiquity proceeded fairly quickly, and the religion's success was far more comprehensive there than it was in Western Europe. By late antiquity, Aristotle's works were comparatively little read, though some writers, such as St. Augustine, adapted Aristotelian logic to a viewpoint amenable to Christianity. By around a half millennium after the death of Aristotle, much of what was left of ancient learning was frowned on as "pagan."

After the Arab conquest of the Near East and North Africa in the mid-seventh century, the remaining territory of the Eastern Roman Empire was thoroughly reorganized by Heraclius I (r. 610–641) and his successors. They made Greek the official language, replacing Latin, and effectively made the state Greek in every respect. Known now as the "Byzantine Empire,"[49] the prosperous realm possessed the surviving works of Aristotle and many other ancient writers. Indeed, the major works of Aristotle had been known continuously from the fourth century BC in the Greek-speaking Graeco-Roman world of the eastern Mediterranean, including much of Egypt, the Levant, Asia Minor, and the Balkans. With the formation of a Greek empire, its intellectuals continued to have full access to Aristotelian and other ancient Greek thought throughout the Middle Ages—for *eight more centuries*—without any significant effect on Byzantine science.

[48] Wood (2007); cf. chapters 1 and 6.

[49] In modern historical works. It remained in Greek the "Roman Empire" right down to its end in 1453.

The Byzantine Greeks became more and more rigidly religious. They rejected Aristotle and the other ancients, and increasingly lost what little science and technology they had retained from antiquity; the nascent science of the classical Graeco-Roman world actually declined. The Byzantine Greeks were in large part unable to learn from their Islamic or Western European neighbors, whom they despised, so they also did not acquire the Central Asian recursive argument method or the college, among many other things. Despite at least two "renaissances" of learning, at no time during the period of existence of the Eastern Roman–Byzantine Empire—more than a millennium—did science progress significantly beyond its ancient stage, nor did a full scientific culture complex ever develop there.[50] Byzantine culture stagnated and eventually disappeared.

## Classical Arabic Civilization

The first culture to develop science per se in anything like the modern sense was Classical Arabic civilization. The presence of science is clear from the appearance of mathematically advanced scientific works imbued with the scientific viewpoint and a true scientist's self-consciousness, beginning in the ninth century. However, this science only appeared after the translation of important scientific works into Arabic from two very different cultures: first from the Indian tradition, mainly in the second half of the eighth century; and, beginning at the very end of the eighth century or beginning of the ninth, from the Greek tradition, either via Syriac or directly from Greek, as discussed in chapter 5.

Most of the greatest scientists of Classical Arabic civilization were Central Asians. They retained the Central Asian Buddhists' endowed college, the *vihāra*, which they Islamicized and renamed the *madrasa*, and eventually spread it throughout the Islamic world. Significantly, however, its curriculum was (or soon came to be) based on the study of Islamic subjects, especially Islamic law; other subjects were pursued mostly outside the college. In fact, all of the greatest scientists of Classical Arabic civilization were privately educated and privately supported. Though there were eventually many *madrasas*, they had little or no effect on the development of science in the Islamic world, where science never acquired a secure institutional base.

The Central Asian Muslims also adopted (or rather, continued using) the Central Asian Buddhists' recursive argument method, but like the

[50] See Grant (1996: 186ff.) for a thorough discussion.

Buddhists, the Muslims seem to have used it mainly in works on metaphysics and theology, although Avicenna also uses the method in *al-Nafs* (translated as *De anima* 'On the Soul', or 'Psychology'), including sections of it that deal with natural science, including optics.[51] However, most of the famous works on natural science produced by Classical Arabic scholars seem to be written in Aristotelian treatise format.

Classical Arabic civilization had the idea of science, as well as Aristotle. But it did not fully utilize the endowed college and the recursive argument method, which it had inherited from the Central Asian Buddhists. Although the Classical Arabic scholars developed full science and produced many good scientific works, they did it largely outside of Islamic culture, which never accepted "natural philosophy" except very grudgingly. It is, therefore, not really surprising that Classical Arabic civilization did not develop a full scientific culture complex.[52]

## Medieval Latin Europe

Science next appeared in the thirteenth century in the Medieval Latin culture of Western Europe. The elements that had contributed to the appearance of science in Classical Arabic civilization were more or less simultaneously transmitted to Europe along with the college and the recursive argument method, and this time a full scientific culture complex developed quite rapidly.

The Greek component—including many previously unknown works of Aristotle—was transmitted via Latin translations made directly from Greek as well as translations of Arabic translations of the Greek, interpreted through the great Arabic commentaries on the same works. The new translations of Aristotle, especially his works on logic, metaphysics, and natural science, transmitted the idea of science and also the basis for the independent fields of the sciences. Aristotle's genius was as important as his errors, which stimulated scholars to try and disprove his views. And it was above all the great Classical Arabic commentaries on Aristotle and the many innovative works of science translated from Arabic that had the greatest impact in their Medieval Latin translations.

[51] The *De anima* is a major part of Avicenna's summa, the *Kitāb al-shifā'*, q.v. chapter 5.

[52] Crombie (1953: 11), speaking of the translations into Medieval Latin of Greek and Arabic scientific works, notes, "The new translations, of which the Greek and Arabic originals had so conspicuously failed to produce a thoroughgoing experimental science in the classical and Mohammedan worlds, provided Western Christendom with the beginnings of a method of rational explanation of empirical facts."

Even more important in the long run, two non-Western components of Classical Arabic civilization, the college and the recursive argument method, were transmitted at the same time. The Central Asian Buddhist endowed college was transmitted via the Islamic Near Eastern variant. In Western Europe, the college soon merged with the *universitas*, or corporate guild of scholars, to produce what we know as the university. As in the tenth through twelfth centuries in Classical Arabic civilization, when the *madrasa* spread across the Islamic world, the thirteenth through fifteenth centuries saw the spread of the college-university across Medieval Latin Europe. Before the appearance of the college in these two cultures, one finds in the record mostly only mosque-schools and a few libraries in Islam, and cathedral-schools in Western Europe. The *universitas* guild did not begin to resemble anything like what we think of as a "university" until after the appearance of the college.[53] The main *intellectual* difference between the Islamic and the European college is that the curriculum of the developed Islamic *madrasa* consisted mostly of religious jurisprudence (*fiqh*) and theology,[54] while that of the developed Western European college consisted above all of Aristotelian "natural philosophy" as well as theology. The main *institutional* difference is that, unlike the Islamic *madrasa*, the Western European copy of the latter, the college, was soon joined to a well-developed scholars' guild, the *universitas*. Together, the college-university quickly developed into the actual ancestors of our present universities.

The recursive argument method was transmitted literarily to Western Europe in the mid-twelfth century via the translation into Latin of Avicenna's *Kitāb al-Shifā'*, including the *De anima* and the *Metaphysics*, both of which were very widely read and cited.[55] The highly formalized and complex framework of Avicenna's version of the method provided an overt structure and rigorous "scientific" way of writing not found in any earlier literature in Europe. Because all scholars had the same basic education in "arts"—at the time meaning almost exclusively natural philosophy, most of which was like what we today call the sci-

[53] See chapter 3.

[54] These were considered to be "Islamic sciences" or fields of learning, unlike Aristotelian natural philosophy, which was considered to be a "foreign science."

[55] See chapter 6. The Central Asian recursive argument method may well have been transmitted to Europe orally first, via European contact with the Classical Arabic version of it during the period of the Crusades in Islamic Spain or the Near East, as argued by Makdisi (1981). However, it seems that none of the medieval Europeans who left records of their travels to Islamic lands mention having encountered the method there.

ences—the recursive argument method was used to compose works in all fields, above all natural sciences, and thus became the literary "scientific method." Major scholars, including Albertus Magnus, Thomas Aquinas, and Roger Bacon, produced massive works known as *summas* 'summaries' on metaphysics and theology, commentaries on Aristotle, and other works, written in the striking new recursive argument method format.

It may have been the sudden appearance of so much advanced culture at the same time, transmitted almost entirely from the Islamic world from the mid-twelfth century to the mid-thirteenth century, that overwhelmed the initial opposition to it from the guardians of the earlier medieval tradition, the conservatives of the day. Science was eagerly accepted by nearly everyone, including the Catholic religious authorities. By the middle of the thirteenth century, the new, independently funded, secular university was forming from the combination of the newly introduced endowed college and the existing *universitas* guild of scholars. The curriculum was based on the new translations of Greek and Arabic texts on natural philosophy, which were made the required texts for all university students at the bachelor's degree level,[56] regardless of whether they might go on for advanced study in law, medicine, or theology. Everyone thus studied the same major texts on natural science and metaphysics. Science eventually permeated the entire intellectual culture of Western Europe, which thus developed the world's first full scientific culture by the middle of the thirteenth century. Despite periodic attacks by anti-intellectuals, science continued to prosper and develop in Western culture down to the late Renaissance and Enlightenment, when a number of factors produced the changes of the scientific revolution. From that point it has continued on down to the twenty-first century.

## The Decline of Science in the Medieval Islamic World

Although science, and all of the other individual elements that in Western Europe were to constitute a full scientific culture, already existed for a time within Classical Arabic civilization, they were not integrated together. As already noted, natural philosophy was considered alien and un-Islamic, or even anti-Islamic.[57] By the lifetime of Averroës (Ibn Rushd, 1126–1198), everything not considered "Islamic"—scientists, Aristotelian thought,

[56] The medieval bachelor of arts degree was thus more or less the equivalent in intention to the modern American bachelor of science degree.

[57] Cf. Grant 2007: 328).

and eventually the "foreign sciences" in general—was under pressure in the Islamic world. The pressure came from governmental and religious authorities but, significantly, it was inspired and encouraged by leading anti-intellectual academics, among them the brilliant scholar al-Ghazālī (1058–1111). The success of the antiscientific camp was nearly total.

Recently it has been increasingly claimed that science did not decline in the Islamic world and that scientific works continued to be composed until modern times. Al-Ghazālī is portrayed as a great scientist, and so on. To a sober historian aware of the calamitous history of the Middle East—including North Africa—for the past few centuries this might seem like a surprising claim. However, such a view is supported, rather hesitantly, even by David Lindberg, who says diplomatically, "Perhaps the question that we ought to be asking is not 'Why or when did Islamic science decline?' . . . but 'How is it that an intellectual tradition that began in such unpromising circumstances developed an astonishing scientific tradition that endured as long as it did?' It may be that [the diversity of this] . . . multireligious, multilingual, cosmopolitan culture . . . ensured that there would remain enclaves of educated, theologically tolerant people, where a scientific tradition, foreign in both origin and content, could take root and flourish."[58]

But in saying that Classical Arabic science "endured as long as it did," Lindberg recognizes what those who work on its history are constantly reminded: it eventually ended as a living tradition. It did not really outlive Averroës. The very partial, highly religious or theological "exceptions" to the great decline in the Ottoman Empire and Persia do not change earlier scholars' blunt assessment.[59] The attempt of Nasr[60] to portray the theology, "theosophy," or mysticism that replaced natural philosophy as nevertheless somehow "philosophy" or "science" is full of inadvertent (and unavoidable) references not only to the actual decline and disappearance of genuine natural philosophy but also, perhaps even more important, to the disappearance of the possibility of speculation, of free investigation, which had been possible for Zakariyyā' al-Rāzī (Rhazes), the greatest physician of Classical Arabic civilization. The situation, in the best of circumstances, was grim. King says of scholars in fifteenth-century Egypt, "Certainly their hearts were in the right place, but circumstances beyond

[58] Lindberg (2007: 191).

[59] *Pace* Wisnovsky's negative remarks about previous scholars' distinction of theology from natural philosophy and their conclusion that the latter did not survive (Wisnovsky 2005: 105, 113, 132–133); cf. Druart (2003: 99–100) and El-Rouayheb (2010: 8ff.) for similar views.

[60] Nasr (2006).

their control rendered their efforts little more than an exercise in futility. And although they surely did not realize it, they were contributing to the decline of Islamic science. . . . Not even the two inconsistent values for the *qibla* at Cairo received a comment."[61] The exceptions clearly prove the rule.[62] But how did this happen?

Like Christianity, Islam is monotheistic and based on a holy book that is considered by believers to be the word of God, and therefore True. It was necessary for early Muslim scholars to try to find a way to accomodate Islam to Aristotelian thought, partly using neo-Platonism, just as Christian theologians such as Augustine had earlier tried to do for Christianity. Accordingly, Avicenna, for example, actually argues against Aristotle on some points, most of them rather substantive ones in metaphysics, the natural sciences, or medicine. However, he did not reject Aristotle's system as a whole or his fundamentally scientific approach; he only wanted to *improve* Aristotle from the point of view of Islam.

By contrast, although al-Ghazālī agrees with very many of Avicenna's arguments, to the point that some medieval thinkers even considered him to be a follower of Avicenna,[63] he explicitly rejects Aristotle's general approach, the idea of science: "Let us, then, restrict ourselves to showing the contradictions in the views of their leader, who is the philosopher par excellence and 'the first teacher' . . . namely, Aristotle."[64]

Al-Ghazālī had begun his career as an Aristotelian philosopher in the tradition of Avicenna, and had been quite successful at it. However, in midlife he had a religious experience and rejected not only "philosophy" (i.e., roughly natural science and philosophy), but scientific thought in general, in favor of mystical Sufism.[65]

[61] King (2008: 341).

[62] Cf. Grant (1996: 176–186). For comments on the issue in a broader historical context, see Beckwith (2009).

[63] Goodman (2006: 38). There is a simple explanation for this medieval mistake. It has been established that the famous work *Maqāṣīd al-falāsifa*, attributed to al-Ghazālī from the Middle Ages on, is actually an Arabic translation of the Persian *Dāneshnāmeh* 'Book of Knowledge' (Goodman 2006: 38, citing Achena and Massé 1955: 44), a précis of Avicenna's revision of Aristotle's system. Although it is often said that al-Ghazālī wrote the *Maqāṣīd al-falāsifa* as the first part of a two-volume work on philosophy, the first summarizing the Aristotelians' views, the second, the *Tahāfut al-falāsifa*, refuting them (e.g., Marmura 2000: xvii), in fact al-Ghazālī never mentions or alludes to the *Maqāṣīd al-falāsifa* in the *Tahāfut al-falāsifa* (Marmura 2000: xvii).

[64] Marmura (2000: 4).

[65] It is worth noting that even mystical Sufism was a Central Asian development in Islam. The first historically known mystical Sufi was Abū Yazīd al-Bisṭāmī, who was from western Khurasan. Ritter (2006) remarks, "Abū Yazīd's teacher in Ṣūfism was a mystic who was ignorant of Arabic, by name Abū 'Alī al-Sindī, whom he had to teach the Qur'ān verses necessary for prayer, but who in

More specifically, in his writings he rejects Aristotle's use of logical and mathematical approaches in fields in which, al-Ghazālī proclaims, logic and mathematics are useless.[66] He concedes that these approaches are needed for the physical-mathematical sciences themselves, and does not contest philosophers' views on such questions per se, but he specifically denies the applicability of such "mathematical" thinking to other fields: "We say: 'As regards . . . the statement that the understanding of metaphysics is in need of [mathematics,] it is nonsense.' It is as if one were to say that medicine, grammar, and philology require it."[67]

Of course *scientific* medicine, grammar (i.e., linguistics), and philology absolutely do require mathematics, as well as logic. Yet al-Ghazālī goes further: "Those who devote themselves eagerly to the mathematical sciences [arithmetic, astronomy, and geometry] ought to be restrained. Even if their subject matter is not relevant to religion, yet, since they belong to the foundations of the philosophical sciences, the student is infected with the evil and corruption of the philosophers. Few there are who devote themselves to this study without being stripped of religion and having the bridle of godly fear removed from their heads."[68] In the case of medicine, and "natural science or physics," in which he includes what we would call physics (including astrophysics), chemistry, and biology, he claims that it is not necessary to reject them except for the objections he enumerates in his famous book *Tahāfut al-falāsifa* 'The Incoherence of the Philosophers', adding, "The basis of all these objections is the recognition that nature is in subjection to God most high, not acting of itself but serving as an instrument in the hands of its Creator. Sun and moon, stars and elements, are in subjugation to his command. There is none of them whose activity is produced by or proceeds from its own essence."[69] He does state that formal logic is necessary for the defense of religion against unbelievers, but he argues that the philosophers have misused it. He thinks it should be used *against* the philosophers, and proceeds to do so in his book to great

return introduced him to the Unio Mystica. It is not impossible that Indian influences may have affected Abū Yazīd through him." Central Asia has continued to be a stronghold of Sufism down to modern times.

[66] Cf. Marmura (2000: xxii).

[67] Marmura (2000: 8). In another work, while building up the same argument on the basis of reasonable statements before making his claim about the inapplicability of scientific thinking to nonmathematical fields, al-Ghazālī says reasonably, "It is not necessary that the man who excels in law and theology should excel in medicine, nor that the man who is ignorant of intellectual speculations should be ignorant of grammar" (Watt 1953: 33).

[68] Watt (1953: 34); cf. Grant (1996: 180).

[69] Watt (1953: 37).

effect. His argument is that if he can disprove their views on metaphysics using logic, their own tool, philosophy as a whole—that is, philosophy and science—is disproved and orthodox religious belief is confirmed.[70] This is his express purpose in writing *The Incoherence of the Philosophers*, which he concludes with the statement: "If someone says: 'You have explained the doctrines of these [philosophers]; do you then say conclusively that they are infidels and that the killing of those who uphold their beliefs is obligatory?' we say: Pronouncing them infidels is necessary in three questions."[71] As for these three religious doctrines, belief in which he declares merits death, he says: "One of them is the question of the eternity of the world and their claim that all substances are eternal; the second is their claim that God's omniscience does not include that of the 'accidentals' of individual [things or people]; the third is in their denial of the resurrection of [people's] bodies and their assembly [at the Last Judgement]."[72]

Al-Ghazālī's "assault against the philosophers"[73] found a ready audience: his book is one of the most successful works about natural philosophy ever written in Arabic.[74] The assault on Avicenna and the philosopher-scientists was thus led by a prominent academic, a former professor of the Niẓāmiyya Madrasa in Baghdad, the equivalent at that time of today's University of Oxford or Harvard University. But if many people of his time and afterward had not been eager to put his recommended inquisition into practice, would he have achieved such success? Surely a single book, no matter how brilliant, cannot bring down an entire civilization.

And indeed, he was not alone. Science had not yet matured intellectually in Classical Arabic civilization when it already came under attack

[70] Marmura (2000: 9).

[71] Marmura (2000: 226).

[72] Text from Marmura (2000: 226), q.v. for his translation. Marmura (2000: xxii) notes that the three doctrines al-Ghazālī condemns, whose supporters should be punished by death, are not trivial; they are fundamental points.

[73] Grant (2007: 87ff.).

[74] Recently, it has been popular to claim that al-Ghazālī was actually an heir of the classical Pyrrhonists (the ancient "Sceptics") and, like them, he was arguing against the dogmatic Aristotelians. Floridi (2002: 22) refers to al-Ghazālī as "the Arabic Hume," and further suggests that al-Ghazālī was a Sceptic (Floridi 2010: 274). This approach completely ignores al-Ghazālī's actual views, which he expresses openly in the *Tahāfut al-falāsifa* and other well-known books, all of which make it abundantly clear that he was a devoted but dogmatic Muslim with a strong fundamentalist streak; cf. Marmura (2000: xvi, xxiiiff.). That several times in the *Tahāfut al-falāsifa* al-Ghazālī uses arguments against his opponents even though he personally does not hold the positions he argues shows merely that he is pragmatic when arguing, as he himself says. He expresses his own dogmatic views throughout the same work. He was no Sceptic, not to speak of a Pyrrhonist of any conceivable kind. For Pyrrho's thought, see Beckwith (2011).

from dogmatic theologians. Aristotle and the idea of science, specifically, were the main targets. Scientists in the Islamic world were persecuted ever more aggressively as time went on. Once Zakariyyā' al-Rāzī (Rhazes) had debated rather freely about much that was considered heretical or anti-Islamic in his day, but a century later—in the golden age of Classical Arabic civilization—Avicenna, Ibn al-Ḥaytham (Alhazen), and al-Bīrūnī were all imprisoned at one time or another and important scientific works were destroyed.

There were many other tracts written against science, or "philosophy," before and after al-Ghazālī, but his *Tahāfut al-falāsifa* cast the stone that started an avalanche.[75] According to its intent, its title would be more accurately rendered into Modern English, following the Medieval Latin interpretation, as 'The Destruction of Science'.

By the following century, in the time of Ibn Rushd (Averroës), science was under open attack and in rapid decline throughout the Islamic world. In his *Tahāfut al-tahāfut* 'The Incoherence of The Incoherence',[76] Ibn Rushd refuted most of al-Ghazālī's arguments, but his works did not have their intended effect locally, and did not make their way east in time to have any effect in the central Islamic lands either. He was too late. His *Tahāfut al-tahāfut* did not succeed in undoing the damage wrought by al-Ghazālī. Philosophy and science came to be considered heretical, and eventually anything that even suggested them was suppressed, taking Classical Arabic intellectual civilization as a whole along with it.

The Islamic world thenceforth slipped increasingly into religious bigotry and scientific and technological backwardness. The victory of al-Ghazālī and his allies was complete by the time the Portuguese learned how to sail around Africa to the Orient, threatening the Levantine Muslims' control over the lucrative spice trade via the Indian Ocean. Large, powerful Muslim states in the region, including even the Ottoman Empire, then at its height, responded by sending fleets of warships against the Europeans via the Red Sea and Persian Gulf, then the "back doors" of the Islamic Middle East. But the Muslims' ships were old-fashioned, and furnished with old-fashioned weapons. Their attempts to restore their previous monopoly on trade with India and the Far East were easily defeated by the Portuguese, who came from a tiny, poor country that was very far away by sea.[77]

[75] The book was translated into Latin in 1328 as *Destructio philosophorum* (Grant 1974: 812), or *Destruction of the Philosopher-Scientists.*

[76] Or "*The Destruction of* The Destruction."

[77] Beckwith (2009: 211ff.).

The Portuguese and their European successors observed and investigated the new cultures they encountered from the very first day they reached India, as we know from the detailed diary of Vasco da Gama and documents written by later explorers. The Europeans' passion for science—a passion to learn about all aspects of the world, to record what they learned, and to improve upon earlier knowledge, which marks them from their very first voyages of discovery—had once been characteristic of Classical Arabic civilization too, but by the time of the European "Age of Exploration" that passion had become exceedingly rare in Asian cultures, whose men of learning almost completely ignored the Europeans and therefore learned practically nothing from them.[78] Although generally overlooked, there was also a Renaissance in Asia,[79] but unlike the Renaissance in Europe it was not the beginning of a great revival and expansion of scientific learning. The reason for this difference is surely to be found in the different attitudes toward learning and, especially, science. Depending on the individual Asian culture, either science had been largely suppressed (as in Islamic culture) or it had never really existed.

One might think that when the college appeared in Classical Arabic civilization, some individual scientific thinkers received support for their work from society, thus contributing to the growth of a scientific culture. But in fact, as noted above, none of the great Classical Arabic scientists studied or taught in colleges,[80] with the one significant exception of al-Ghazālī, who was educated in *madrasas* in Ṭūs, Jurjān, and Nishapur, and taught for some years at the great Niẓāmiyya *madrasa* in Baghdad,

[78] A number of scholars have argued that such views are incorrect because there do exist travel accounts written by Muslim visitors to Europe in the seventeenth century (Matar 2003, 2009). However, these were the exceptions that prove the rule. The most prolific and polished of these writers, Evliya Çelebi, traveled in the Islamic lands of the Middle East and in the Islamic or partly Islamic lands of Eastern Europe. The other accounts are interesting and certainly important, but they are pale reflections of the intellectual brilliance of Classical Arabic civilization at its height, and are in no way to be compared to the vast, detailed literature on non-European peoples produced by Western Europeans not only in the seventeenth century but from the beginning of the Age of Exploration from the end of the fifteenth century to the modern period, and still continuing. One of the most remarkable things about al-Bīrūnī's *India* is the fact that the book was written and still exists. There is nothing similar to it in Arabic literature, which is not even remotely comparable in this respect to Western European literature from at least the fifteenth century on. This radical difference needs to be studied and explained, not "explained away."

[79] On the Renaissance in Asia, see Beckwith (2009), where, unfortunately, it is not pointed out that the Renaissance there was a last gasp and was marked primarily by perfection in the arts, which thereafter also declined precipitously.

[80] Cf. Grant (1996: 178).

where he wrote his *Tahāfut al-falāsifa* and several related works.[81] In light of the sharply differing history of medieval Western Europe, the failure of Classical Arabic civilization to build a lasting intellectual culture—and the success, instead, of a strictly religious, increasingly anti-intellectual culture—appears to be due above all to the failure of Islamic society to use the new Central Asian college, or another educational institution, to give science a firm foundation and encourage its growth. Instead, the *madrasa* came to be used largely by theologians and the sectarian partisans of Islamic law, who spread their colleges across the Islamic world and quickly won the battle, and the war. After the victory of religious dogmatism, there was no need for colleges anymore; they began disappearing from most of the Islamic world shortly after the destruction of science there.

[81] Marmura (2000: xvii).

## — Chapter Eight —
# Conclusion

In the twelfth and thirteenth centuries, Europe underwent radical cultural changes that affected most of the major fields of human activity studied by historians and have been referred to collectively as a medieval "intellectual revolution." The origins and consequences of some of these changes for the development of a full scientific culture in Western Europe have been examined above in some detail.

In this chapter, three objections are raised and answered about the development of a full scientific culture, and the modern descendants of the recursive argument method are described and discussed.[1]

### Development of a Full Scientific Culture

1. Why should it have been specifically the combination of two originally related elements, the college and the recursive argument method, that brought about the "quantum leap" from the state of having a few scientists and some good science, as in Classical Arabic civilization, to the state of having a full scientific culture?

2. If civilizations that had full knowledge of ancient Graeco-Roman science did not develop a full scientific tradition because they lacked the college and the recursive argument method, why did some civilizations that *did* have the college and recursive argument method, as well as some science, nevertheless *not* develop a full scientific culture?

3. What is the difference between a civilization that has a few scientists and some science, and a civilization that has a "full scientific culture"?

It has been shown in the preceding chapters that the essential component elements of medieval science all came from outside Latin Europe. They were therefore both alien and discontinuous, and consequently prime candidates to bring about revolutionary change.

[1] This concluding chapter, like the introductory chapter, is an essay, and as such is very lightly annotated. For references, see the detailed discussion of the respective points in the preceding chapters.

One of these essential elements, Aristotelian logic—particularly the previously unknown works constituting the *Logica nova* 'new logic'—has long been considered to be the most important element, if not the only important one, in the development of medieval science. It was introduced through translations into Latin from Greek and Arabic, along with translations of mainly Arabic commentaries.

However, the idea that the works of Aristotle were the most important factor in the establishment of a full scientific culture in medieval Western Europe is not supported by the historical sources. Basic Aristotelian logic had already been known in Latin Europe for many centuries via the *Logica vetus* 'old logic', consisting of Aristotle's *Predicamenta* 'Categories' and *De interpretatione* 'On Interpretation', as well as Porphyry's *Isagoge*, an introduction to Aristotelian logic. They had been translated into Latin in late antiquity by Boethius, some of whose own works on logic were also known throughout the Middle Ages.[2]

Moreover, the *Logica Nova* works of Aristotle, though certainly important, not only had failed to set off a revolution of any kind in the Greek speaking world for the previous millennium and a half, they also had at first virtually no effect on medieval Western Europe when good Latin translations were made directly from the Greek in the early to mid-twelfth century. Scholars had already been prepared, through long acquaintance with Aristotle's way of thinking in the *Logica vetus*, to receive the *Logica nova* and understand it, but they hardly reacted at all. Why then did the very same ancient Greek thought suddenly have a revolutionary effect on Latin Europe in the thirteenth century and become centrally important to the new science?

The answer might be sought in the other salient constituent elements of medieval science, all of which were new and fully or partially external in origin as well, namely: the recursive argument method; the college, which merged with the native *universitas* guild to produce the university during the thirteenth century; Indian place-system numerals and mathematics; and the accumulated science of Classical Arabic civilization, including long-influential works on alchemy, astronomy, mathematics, metaphysics, medicine, physics (especially optics), psychology, and many other fields. Among all these elements, the recursive argument method and the college had previously been inherited or borrowed by Central Asian Muslims from the Central Asian Buddhists, while Indian

[2] Murdoch (1974: 77). Cf. Grant (2007: 108), who notes that the *Logica vetus* "played a negligible role in Western Europe" until the eleventh century.

mathematics had been borrowed and adapted by the great Central Asian Muslim scholar al-Khwārizmī (Algorithmus) and his successors in Classical Arabic civilization, the leading lights of whom were also mostly from Central Asia.

The latter elements are thus not only non-European, they are non-Western. They are exactly the ones that were most alien to Western culture and therefore the most potent agents for change. These elements were so influential even after they had been "Islamicized" that the Latin Europeans were forced to change their own culture in order to accept them. In accepting them from Classical Arabic civilization (along with poetry, music, and much else), Latin Europe underwent an intellectual revolution. In it a scientific culture complex developed, in which science was institutionalized, such that all people educated in universities from then on received their basic training in medieval science. European intellectual culture, thus fundamentally restructured, led eventually to the Renaissance discoveries and subsequently to the scientific revolution of the Enlightenment and on to modern science. In short, it was only after Western Europeans got the idea of what Aristotle was about from the Aristotelian works of Avicenna and other scholars such as al-Fārābī, and from the Classical Arabic commentaries on Aristotle, that they got Aristotle's idea of science. Because the Western Europeans' awakening happened along with the contemporaneous importation of the recursive argument method and the college, all the above elements became bound together in a full scientific culture. Aristotelian science was argued about in the college-universities following a rigorous, transparent argument structure, the recursive argument method, the "scientific method" of medieval science.

In recent history there are better recorded and better known examples of just this sort of revolutionary development. Consider for a moment the Meiji Restoration, the Japanese revolution that began in the mid-nineteenth century when the country, which had been officially sealed off from Western influence for more than two centuries, was forcibly opened by American naval vessels in the 1850s. The Japanese quickly understood the danger they were in and promptly overthrew the aging Tokugawa shogunate. Under the leadership of the restored Meiji emperor, the Japanese adopted Western technology, commercial practices, law, literary forms, architecture, music, sports, and practically everything else, including science and the entire Western education system from primary school through the university. Within less than a half century, the Japanese assimilated this massive foreign influence and their country became a world power. It is true that Japan had already undergone a little Western influence beginning with the Portuguese establishment of commercial relations in the mid-sixteenth century and continuing during the period of isolation via a trickle of books obtained from the Dutch trading post in Nagasaki Bay.[3] This does seem to have helped prepare the Japanese somewhat to deal with the sudden threat. But the scale of the subsequent flood of foreign influence was completely different from what had gone before, and was magnified by its suddenness, simultaneity, and multifaceted nature. The impact of Western culture on Japan in the mid-nineteenth century was truly overwhelming. The result was spectacular.[4]

Similarly, the contact of Western Europe with Islamic culture during the period of the Crusades extended from the frontiers of Christian Spain all the way across Islamic North Africa and the Near East to the Byzantine Empire. This contact was on a massive scale. It is not surprising that it had a revolutionary effect on Medieval Latin civilization. It would be surprising if it had *not* had such an effect.

With these points in mind, replies may be given to the questions raised above.

1. The college and the recursive argument method, though they had developed together in Buddhist Central Asia, and continued to develop in Islamic guise when that region converted to Islam, seem to have been focused mainly on theology and religious jurisprudence in Classical Arabic civilization. It seems likely that because they were transmitted together, at the same time, to Medieval Latin civilization—and very likely because one of the earliest works translated, Avicenna's *De anima*, includes examples of the method's use for topics in natural science—they

[3] There were some transmissions of learning from the Islamic world to Western Europe before the mid-twelfth century—e.g., Gerbert of Aurillac, q.v. Lindberg (2007: 201)—so the two cases are closely parallel.

[4] Presumably, no one has actually proposed that the Japanese independently invented baseball, such that the American development of the game based on English forerunners would have to be seen as a coincidence. But it is easy to find many actual examples of claims like this imaginary one in the historical literature about Medieval Latin civilization. Extremely few scholars actually accord to the influence of Classical Arabic civilization on Medieval Latin Europe the fundamental, paradigm-shifting importance that it deserves. The same is true of scholarly treatment of the early period of Classical Arabic civilization, which has recently become even more parochial and xenophobic in general than it was a few decades ago, so the once-noted powerful influence of Central Asia and India on the formation of Classical Arabic intellectual civilization *before* the influx of Greek thought is now barely mentioned even in passing, attributed to imaginary Persian intermediaries, or simply ignored. For example, contrast Pines (1936) with Van Ess (2002). The parochial shift can even be seen in successive works on the same subject by the very same Islamicist scholar, as noted in Beckwith (1984b).

were seen as important elements of the new way of thinking that came in along with all the other elements. The "quantum leap" from the state of having a few scientists and some science to the state of having a full scientific culture thus happened because of simultaneity and synergy, supported powerfully by the immense prestige of anything that came from Classical Arabic civilization and was therefore more "advanced" than what was already known locally. In other words, the Islamic "intellectual revolution" in medieval Western Europe happened more or less as the Western "cultural revolution" did in nineteenth-century Japan.

2. It may be argued that some civilizations, which had the college and recursive argument method as well as some science, nonetheless did not develop a full scientific culture because they did not also know the works of Aristotle, in which the idea of science and the scientist is very clearly developed. However, this is not supported by the historical evidence. As noted above, Aristotelian science was already known (though, to be sure, not all of it) in earlier medieval Europe, yet science per se hardly developed there, not to speak of a full scientific culture. Even more striking is the fact that a full scientific culture never developed in the Graeco-Roman world or its Byzantine continuation despite an unbroken millennium and a half of more or less full knowledge of the works of Aristotle and his rivals—and what science survived there even declined. Finally, it is significant that although advanced science did develop in Classical Arabic civilization, it eventually declined or was suppressed more or less completely because science was never integrated into the dominant Islamic culture. Even though the college and the recursive argument method were known and used, to some extent, in that civilization, they were not integrated with science, which therefore had no institutional and methodological base. There was also no other regular educational system devoted to natural philosophy that passed on its traditions to later generations. Few of the great scientists of Classical Arabic civilization used the recursive argument method in their works, and none were educated in a *madrasa*—al-Ghazālī being the putative exception that proves the rule. The difference between the fate of these cultural elements in the medieval Islamic world and in medieval Western Europe is striking.

3. The difference between a civilization that has a few scientists and some science, and a civilization that has a full scientific culture is evident not from the study of fields that contemporary twenty-first-century scholars think are "scientific" but from the study of other fields *not* commonly thought of as being "scientific." It is the treatment of such fields as sciences in Western Europe from the thirteenth to the twentieth

centuries that allows that culture to be characterized as having been fully scientific.[5]

Here it is necessary to point out the unpleasant and much ignored historical fact that from the early twentieth century on, Western civilization has undergone radical, deep changes that have destroyed many long-standing cultural traditions that once characterized that civilization. Their source is a little-studied movement that has been labeled "Modernism."[6] Historians have hardly understood this movement, which was the driving force behind the incredible destruction of the two World Wars, other devastating wars in the same period, mass murder unparalleled in world history, the death of the traditional arts, and so on. What little of Modernism is even recognized at all by historians is mostly praised, as if not enough has been destroyed yet.[7] Because Modernism remains fundamentally unstudied and unknown, its historical effects remain mysterious, and its ongoing destruction of cultural traditions remains little noted. Science has not miraculously escaped. It is under continuous attack from all sides, including from academia, and has survived to the present only because of its inextricable connection to technology, without which modern civilization would collapse utterly and many profitable businesses would go bankrupt. Modernism continues its unbridled course of destruction and antipathy toward anything that smacks of science outside of technologically critical fields.

Now consider traditional pre-twentieth-century Western European culture, in which a "scientific" approach to everything was still its most striking characteristic. It is not just that fictional characters such as Sherlock Holmes are successful and praiseworthy because of their use of logic and scientific methods to solve problems. Everyone, in all fields, from art and music to history and philology, was still educated first in the sciences, and they referred to their own way of working as "scientific," as indeed it was. This was not just odd for observers from other cultures; it was incomprehensible. What could science or a scientific approach teach anyone about art or ancient philosophy?

[5] Eduard Vilella (personal communication, December 20, 2011) notes that scholarship on the invention of the sonnet in Sicily during the time of Frederick II, when Sicily was one of the centers of Islamic-to-European culture transfer, has established that its structure follows the recursive argument method and the poems are presented as arguments. See Arqués and Pinto (1995). I am of course responsible for any misunderstanding of his remarks.

[6] See Beckwith (2009), where the discussion focuses mainly on political history and the arts.

[7] Modernism has nothing to do with what might be called "technological modernity," a "steampunk" sort of view of the artifacts of the recent past that glorifies it, or with general "modernity."

From at least Aristotle on, the meaning of science has been partly sociologically determined, in that scientists' self-consciousness of the difference between science and nonscience, and between scientists and nonscientists, is a fundamental element in a culture that has developed advanced science. This is openly stated by the scientists of Classical Arabic civilization, perhaps partly because they were under constant, increasing pressure from those trying to prevent them from doing science. Scientists are thus consciously aware of the difference between science and nonscience. They are aware that they are scientists, they are aware of the fact that they are working scientifically toward the goal of producing scientific results, they are aware that the goal of science is to find the best possible answers to the questions they ask, and they are acutely aware of the existence of many who do not understand science or are even opposed to it. This self-recognition is an absolutely central feature of science; nonscientists necessarily cannot include this particular reflective view in their own self-definitions, at least not in the way scientists can and do.

In this connection it must also be emphasized that the specific subject matter of scientific research is, in principle, irrelevant.[8] Scientists can do research and write scientifically on whatever topic they wish, as long as it is amenable to being examined with scientific methodology. The great Central Asian scientist al-Bīrūnī considered himself a scientist. He wrote many important works of science, including the first ethnography of India, which he considered to be a scientific work. This does not need to be inferred. He explicitly says so repeatedly in the book, in which he emphasizes the difference between what is scientific and what is nonscientific. Other scientists from the Classical Arabic and Medieval Latin civilizations are equally explicit—and emphatic—about this issue.

People of all times and places have wondered about the universe and have occasionally written down their musings, their attempts to figure out a perceived problem, their observations of this or that natural phenomenon, their mathematical calculation of this or that. Sometimes they have applied their practical abilities, combined with their understanding of natural physics, to problems of daily life, especially engineering problems, with more or less success, producing technological "progress." The popular use of the word *science* today thus includes contributions

[8] This is comparable to the situation with the recursive argument method, which by itself is also content-neutral. It would seem likely that this flexibility as a tool for scientific examination of questions is what gave it the aura of science.

considered to be "scientific" *today*—especially in mathematics, the natural sciences, and technology—that are actually found within works of any kind, including modern ones as well as those written in ancient Greek, Latin, Chinese, Sanskrit, and so on.[9]

Such early scientific discoveries are in most cases buried within otherwise nonscientific texts, as al-Bīrūnī acidly remarks about Indian science.[10] In modern popular usage, therefore, the word *science* in this sense includes the products of a fully scientific culture as well as the products of fully nonscientific cultures and everything in between, made by people working according to any methodology or no methodology at any time in any kind of culture, and presented in any way in literature or archaeological artifacts. The only criterion for them (and their cultures) to be considered "scientific" is that a *modern* author has declared them to be so. The many histories that treat "science" in Babylonia or premodern China side by side with that in late medieval or early modern Europe, for example, necessarily use the word *science* in this way. Such mixing of the products of radically different kinds of cultures is done by modern scholars who belong at least tangentially to what is left of the European scientific culture complex. They know what scientific ideas are in their own scientific tradition—or rather, what scientific *products* are in their own increasingly unscientific or antiscientific cultural tradition—and they simply extend this modern European definition of *science* to premodern nonscientific cultures. In fact, whatever a discovery is, it is only "scientific" when it has been identified as "scientific" by those belonging to a scientific culture.[11] With extremely few exceptions, use of the word *science* for the products of premodern non-Western cultures requires the ahistorical application of specifically Western concepts to them.[12] In short, science per se exists only among people who have the *idea* of science, though their culture should also have scientific institutions, among other things.[13] But whatever *word* such people use for it is irrelevant as

[9] E.g., Serres (1995).

[10] See the quotation in chapter 7.

[11] Cf. Gower (1997: 251): "It is after all we . . . who label some part of this work 'scientific'."

[12] This is, of course, cultural imperialism if there ever was any. It is one of the biggest (and commonist) sins of historical scholarship to impose the categories of one culture on another—i.e., to think that because we have cultural feature *x* the foreign historical culture we are studying also must have *x*. For a discussion of an example of this error that continues to fundamentally mar the historiography of East Asia, see Beckwith (2009: 321–362).

[13] The products of scientists' work are not a *defining* feature of science because they constantly change over time, sometimes drastically. This change can occur in the work of one scientist who changes his or her mind on a scientific question through further research or reflective thought, or in the work of a scientific community or an entire field over a long period of time

long as it means to them approximately the same thing as what *science* means to people *today* who live *in a scientific culture*.

As noted above, scientists are aware of the difference between themselves on the one hand and nonscientists in the same culture or other cultures on the other hand, with respect to the distinction between a scientific way of thinking and working on problems versus other approaches. By contrast, in nonscientific cultures there are no scientists per se, and there is no scientific context; strictly speaking, science does not exist there. Historically, therefore, the "scientific" products of a nonscientific culture are completely different from the products of a culture that has modern science.

This may be illustrated by an example. Penicillin was discovered, it is said, by accident—it was found in an ordinary mold that had contaminated a bacterium culture. But in a nonscientific culture it is hardly conceivable that penicillin would have been discovered in this way, or probably in any way, because a nonscientific culture does not have the specifically scientific mindset, methodology, and procedures involved—in this case, the practice of growing bacteria in a controlled environment as part of an experimental study of "the role of antibacterial agents in the body and in the natural defense against disease"—which led Alexander Fleming to make the discovery of the "growth inhibition of *Staphylococcus* by fortuitous contamination with the mold *Penicillium notatum*," which he reported in 1929.[14] Penicillin thus was *not* in fact "accidentally" discovered, nor was it "coincidental" that several other scientists doing the same kind of research within the same cultural paradigm—the modern scientific culture complex—discovered essentially the same thing. There are countless other examples that show that the products of modern science are inseparable from a full scientific culture.

Science can permeate the intellectual culture of a society, as it did in Europe from the thirteenth century up into the twentieth century; or it can exist as a subculture within the intellectual culture as a whole in a society, as it did in Classical Arabic civilization, and as it increasingly does today in many "developed" countries. This is especially noticeable in countries where the natural sciences have only recently been introduced by contact with European cultures. The difference between whole-culture approval on the one hand and toleration as a subculture on the other does not seem to affect scientific output very much in either case today, but it does affect the development of intellectual life in

[14] Summers (2003).

general, and can even affect the development of society as a whole, in a given culture.[15]

This may be best understood from another example. Many historians are aware of Renaissance humanists' attempts at manuscript edition, but few, perhaps, are aware of the later development of the science of critical text edition—and if they were, they would probably not call it science unless they went to the trouble to try and became informed about it. Nevertheless, by the nineteenth century, philologists—the same scholars who followed strictly scientific principles to found the field of linguistics—had developed critical text edition into a scientific methodology, which they used for texts in ancient Greek, Hebrew, and Latin, as well as medieval texts in Greek, Latin, Arabic, Old French, Old English, and other languages.[16] From doing many such editions and dealing with the problems involved, they developed a theory of critical text edition. The field became fully scientific in every significant respect. Although nowadays its practitioners themselves rarely call what they are doing "science"—mainly because of the ongoing Modernist reaction against science—they do distinguish their work very sharply from what is not done, or is done badly, in some fields unaquainted with critical text edition.

For instance, in the field of East Asian studies even the term *critical text edition* is unknown, and the methods and theory of textual scholars are either unknown or misunderstood.

> Sinology in this century has not risen to the recurrent challenge to rectify texts. In the other branches of textual scholarship, that is to say in epigraphy and the history of texts, we have, it is true, made some indecisive advances, but in textual criticism we seem, by and large, to have been content to rest on the achievements of [native Chinese scholars of] the eighteenth and nineteenth centuries. This failure is particularly conspicuous when one considers that the Europeanists, both classical and vernacular, have been prodigiously active in the practical tasks of editing and, during the past twenty or thirty years, in the development of the theory of textual analysis. . . . [W]e cannot continue simply

[15] When a society as a whole, or even just its general intellectual culture, is strongly opposed to science, there is a great possibility that it can be suppressed. See above on the fate of science in the Islamic world.

[16] Shapin (1996: 75ff.) refers to the very beginnings of this movement, "humanism," in which the scholars' interest was mostly in attempting to recover the "original pure and accurate" texts of classical works. The modern meaning of the term *humanism* as, in effect, a branch of scholarship characterized by the *lack* of scientific methods, is thus virtually the opposite of its meaning in the Renaissance.

> to ignore the fact that all ancient Chinese texts at least must be re-edited. Unfortunately the inescapability of this daunting challenge has not strengthened the will of Sinologists to accept it. . . . Until we have mended our texts and trimmed them well, no amount of aesthetic intuition or statistical method will move the study of Chinese antiquity out of the doldrums in which it has languished so long.[17]

The same point was made by Paul Pelliot a century ago, after his visit to Tun-huang, the westernmost Chinese town on the frontier of East Turkistan and the Silk Road.[18] He remarked, "Aujourd'hui nous nous apercevons que la tradition manuscrite ou imprimée n'a pas été impeccable, et qu'il faut faire, en chinois comme ailleurs, de la critique de textes."[19] The scholarly Sinological world was thus told of the importance of doing critical editions of Chinese texts by a scholar who was already one of the world's leading Sinologists. In those days it was still a very small field, so most of them must have read Pelliot's comment within a short time after it was published. Although the field of Sinology has thus shirked its scientific duty for an entire century, there seems to be a fairly simple reason why this failure has been overlooked.

Science was accepted by East Asians themselves quite late, and at first only as a necessity, insofar as it could help them to achieve military (and thus political) parity with the West. Other than in the physical sciences (including medicine) and related technology, science was soundly rejected. The native scholars of East Asia considered their traditional interpretations of their own history, literature, and so on to be far

[17] Thompson (1979: xvii). Unfortunately, perhaps partly because Thompson's work is an edition and reconstruction of fragments quoted in other works, it includes some infelicitous innovations that prevent it from being used as a model for others to go and do likewise. Nevertheless, such failings are not faults of critical edition *itself*, and do not obviate the need for scientific critical editions to be done for all important premodern Chinese texts. Certainly they do not excuse Sinologists, Japanologists, and Koreanists for not bothering to learn and use even the most basic philological methodogy. For easily accessible examples of recent critical editions of major Greek and Latin texts, see the Budé series published in Paris and, in particular, the series' updated guide to the practical production of critical editions of Greek texts (Association Guillaume Budé 1972). Like other such handbooks, it does not include an actual example of texts with a critical apparatus, for which one must examine the actual editions published in the Budé series.

[18] He is famous for his brilliant scholarship as well as for having brought back a large number of manuscripts in many languages, especially Chinese and Tibetan, to study in Paris.

[19] The quotation in Duyvendak's (2001: xx) necrology does not cite the exact reference for Pelliot's comment, but it is clear that it must have been written not long after Pelliot's departure from Tun-huang and arrival in Peking in 1908. Unfortunately, Pelliot did not himself produce a critical edition, which might have provided a model for less enlightened Sinologists to follow.

superior to foreigners' views, so they had nothing to learn from the West. The East Asians' Western students have adopted these opinions enthusiastically, and consider them to be models of enlightenment. The rejection of philological science in Sinology meant the rejection of precise, critical thinking about many subfields of East Asian studies, such as history and linguistics—which field includes comparative-historical linguistics, typology, and even simple linguistic description of major modern languages such as Chinese and Japanese.[20]

In short, in these fields the rejection of science by East Asianist scholars from East and West is as complete as could ever have been wished for by al-Ghazālī, to the extent that it is currently impossible to talk intelligently to anyone exclusively in Chinese about scientific critical edition of texts. This is so not merely because of the nonexistence of good examples to serve as models, but because there is no word or expression in Chinese that means 'critical edition' *in the scientific sense*. The same is true of Japanese and Korean. These cultures are therefore still fully nonscientific with respect to text philology,[21] and the corresponding fields of Sinology, Japanology, and Koreanology are similarly marked by the absence of any knowledge of critical text edition. The crucial distinction is accordingly between a scientific field with a theory and method, on the one hand, and a nonscientific field, on the other.[22] Historically, this distinction applies to cultures as a whole.

Medieval Latin Europe is the first civilization to have developed a full scientific culture. It appeared there in the thirteenth century. Because no new elements that are *essential* for a full scientific culture have appeared in European cultures or elsewhere since that time—despite much changing of emphases, cosmologies, instrumentation, philosophical justifications or attacks, such as those that characterize the scientific revolution of the late Renaissance and Enlightenment—it may be concluded, in agreement with earlier scholars, that medieval science was the ancestor of modern science. Moreover, because the two essential elements of the full scientific culture of Western Europe were imported from Classical Arabic civilization, which in turn had adopted and adapted them from

[20] See the comments in Beckwith (2007a, 2007b, 2010a).

[21] Recently two colleagues, of whom one is American and the other Chinese, have told me that they intend to produce critical editions of the Classical Chinese historical texts they are working on, so perhaps the situation will begin changing for the better soon.

[22] The content involved is literature, which is otherwise not typically thought of as being the subject matter for a scientific field, but this merely confirms that the use of scientific methods does not depend on content.

Buddhist Central Asian institutions, the development of Medieval Latin scientific culture, one of the greatest achievements of the Middle Ages, owes its existence and lasting success to those distant, much ignored people of Central Asia and their once brilliant civilization.[23]

## Modern Heirs of the Recursive Argument Method

In medieval and Renaissance times, all scholars studied all fields of knowledge, at least to some extent, so the recursive argument method was used in more or less all fields of thought, though it was the dominant literary form for works on science.

It seems to be generally assumed that the recursive argument method was abandoned and entirely disappeared in the late Middle Ages, but this is incorrect. As Grant shows,[24] the traditional recursive argument method continued to be widely used, especially in works on science, metaphysical philosophy, and theology, throughout the late Middle Ages and the Renaissance, into the seventeenth century.[25] Nevertheless, after centuries in which science was dominated by the recursive argument method, it came to be seen as an old-fashioned Aristotelian artifact because the *content* of works written according to it was overwhelmingly Aristotelian in inspiration. The method was thus abandoned in its traditional form in the course of the scientific revolution, and largely replaced by the treatise format for more extensive works on scientific topics.[26]

However, a new method, focused on experimentation rather than logical argumentation, developed to take its place as the practical-theoretical underpinning of science. This is the "method of hypotheses and experimentation," or what is popularly called the "scientific method." The method is recursive and thus structurally related to the medieval recursive argument method, on which it seems in part to have been consciously or unconsciously modeled, though the recursiveness is typically not fully overt and constitutes only a small part of a scientific work that

[23] For a short introduction to Central Asian science, see Starr (2009). For more on the historical context, see Beckwith (2009).

[24] Grant (2007: 183–187) translates, in full, the recursive argument of John Buridan (ca. 1300–1358) "On the Possibility of Other Worlds."

[25] Grant (2007: 288; 1996: 199). The recursive argument method was still used in the sixteenth and seventeenth centuries in works written by "natural philosophers" who proposed new, non-Aristotelian solutions to the old questions about nature.

[26] Grant (2007: 288–289). Ironically, the Enlightenment "natural philosophers" rejected Aristotle, who did not know the recursive argument method, but they adopted the treatise format, which he used for practically all of his surviving works.

uses it. This ideal "scientific method" is still widely taught to beginning science students and remains in frequent use for actual scientific works, mainly published reports of laboratory research, which constitute one of the major bodies of modern scientific publications, although most such works consist largely of treatise-structure text.[27]

According to the "ideal" experimental method, the scientist first formulates precise hypotheses (based on earlier ideas and observations), which are explicitly stated; second, the scientist describes exactly how the hypotheses were tested using the techniques and tools of the particular science involved, and gives the results of the actual tests; third, the scientist confronts the hypotheses one by one with the data and results of the tests performed and draws a conclusion about the validity of each hypothesis, in order.

This ideal literary scientific method, the continuation of the recursive argument method, may be termed the *recursive scientific method*. It is still often described in science textbooks, but because there is no longer general agreement among historians and philosophers of science on almost any aspect of science, including its history and its philosophical grounding and practice, not to speak of theories of a scientific method, or even the meaning of *science* itself, it is not surprising that there is no longer a clear model for writing scientific works in most fields, with a few exceptions, and that there now are many literary models for them.

Nevertheless, the overt recursive scientific method is demonstrably in active use and is even required to be used in some scientific publications. Over the course of the twentieth century, the American Psychological Association produced many editions of its *Publication Manual*, which is intended to explain to scientists exactly how to design and write a research article for one of the Association's numerous journals. The *Manual*'s authors claim, "Its style requirements are based on the existing scientific literature rather than imposed on the literature."[28] The *Manual* has been adopted by many other journals as well,[29] and continues to be revised regularly down to the present. Its format remains the standard in experimental laboratory psychology, and is taught to university students specializing in experimental psychology.[30]

[27] See below for an example and discussion.

[28] American Psychological Association (1983: 10).

[29] American Psychological Association (1983: 9). Although many scientific fields do not specify a particular style or format for their articles, they do in fact have one by implication. Authors simply copy the format of a similar article that has already been published in the target journal. Articles in the same journal tend to be quite similar in structure.

[30] E.g., Goodwin (1998).

For scientists who do use the recursive scientific method in their publications, the same basic structure is followed, with very few deviations. There are several categories of articles described in the *Publication Manual* of the American Psychological Association. The following summary[31] describes how their "laboratory report" type of article should be written.

### *The Recursive Scientific Method in Modern Science*

A. The author[32] must give a succinct title in which the Topic of the paper is clearly stated.

B. The Topic is clarified and amplified slightly in the Abstract, which follows immediately after it.

C. The Introduction section

   (a) describes the problem under examination and gives the current state of knowledge about it and published scientific literature relevant to it;[33] and

   (b) states the goals of the study, including any hypothesis or hypotheses tested.

D. The Method section explains the actual research procedure, including description of the "participants" (humans or animals); the apparatus or other research tools (such as tests) utilized; and the design of the study, including controls, the details of how the laboratory work was actually carried out, and what happened to the participants.

E. The Results section gives a succinct verbal description of the experiments' results, along with the most relevant parts of the actual data, including statistics, tables, figures, discussion, etc.

F. The General Discussion section

   (a) summarizes the main results of the research project with respect specifically to the original hypothesis or hypotheses and other explanations mentioned in the Introduction; and it

   (b) draws a conclusion, or conclusions (whether positive, negative, or indeterminate), and suggests directions for further work.

[31] This summary is based on the American Psychological Association's (1983) *Publication Manual* and Goodwin's (1998: 415–423) very similar analysis and recommendations. I have omitted all discussion of purely technical questions of publishing, such as journal editing procedures, formatting of manuscripts, proofreading, and so forth, which are discussed in great detail in these works.

[32] I realize that most scientific papers list multiple authors (even though in fact only one or two of them usually do the actual research and write the paper), but for simplicity's sake I refer to the authors as a corporate body, in the singular: "the author" or "the writer."

[33] The bibliographical references are cited in full at the end of the article.

Based on its actual contents (which do not agree very well with the description of its structure), and on actual examples taken from recent journal issues, the above method can be reanalyzed and presented according to the style used in the present book, as follows:

### *Modern Recursive Scientific Method Argument Structure*

I. Main Argument (Topic or Question [A, B, C (a)])[34]

II. Subarguments$_1$ [C (b), presentation of the Hypotheses]

[1] Hypothesis
[2] Hypothesis
[3] Hypothesis

III. Author's View Argument—Exposition, experiments [D, E—the main part of the work]

IV. Subarguments$_2$ [F (a), evaluation of the Hypotheses in II]

[1] Hypothesis evaluation
[2] Hypothesis evaluation
[3] Hypothesis evaluation

V. Author's View Argument—Conclusion [= F (b)]

Much as in the premodern recursive argument method, wherein Subarguments$_1$ with which the author agrees are not mentioned or discussed in the Subarguments$_2$ section, in the modern scholastic-scientific method it is common for views of the previous literature to be omitted in the later discussion. The literature is analyzed as part of the Topic section because it is presented in the Introductory section and is very clearly distinct formally from the presentation of the hypotheses—the Subarguments$_1$—which are regularly replied to in the Subarguments$_2$.

In order to see how the method works in practice, a recent article from a scientific journal, *Experimental Psychology*, is analyzed in Example 8.1. Note the author's use of explicit numbering for the two main subarguments—the hypotheses—given in both subargument lists. As in the medieval recursive argument method, numbering is also used for other parts of the argument.

[34] The capital letters here refer to the sections of the argument as analyzed above on the basis of the required format specified in the *Publication Manual* of the American Psychological Association, as presented in Goodwin (1998).

Research report–type articles in other fields (including pharmacology and neuroscience) were also examined and found to typically contain reference to hypotheses or assumptions and the results of testing the hypotheses or assumptions, sometimes as explicitly and clearly as in EXAMPLE 8.1, but often far less explicitly and clearly. The only thing that reflects a dim memory of the old recursive argument method in such works is the persistent idea of the desirability of presenting one's hypotheses or assumptions and then giving the results of the tests to which they have been subjected, in order.[35]

### *Example 8.1*

**AMIT ALMOR**

***Experimental Psychology***

Why Does Language Interfere with Vision-Based Tasks?[36]

[I. MAIN ARGUMENT (TOPIC OR QUESTION)]

Why does language interfere with vision-based tasks? Conversation with a remote person can interfere with performing vision-based tasks . . .

[II. SUBARGUMENTS$_1$ (the HYPOTHESES)]

[1] Thus, the *first* specific hypothesis tested here is that the interference posed by language on other tasks will be higher during preparing to speak . . .
[2] The *second* specific hypothesis examined here is that keeping track of a listener's position is an obligatory mechanism that operates even in remote communication, such that . . .

[III. AUTHOR'S VIEW (EXPOSITION—the experiments; main body of the paper)]

[IV. SUBARGUMENTS$_2$ (results of testing the HYPOTHESES)]

[1] . . . This supports the *first* hypothesis tested in this paper, which was that, during planning, speakers utilize . . .
[2] The *second* hypothesis tested in this paper was that spatial attention resources are used . . .

[35] This would be a good topic for a doctoral dissertation.
[36] Almor (2008).

[V. AUTHOR'S VIEW (CONCLUSION)]

> It should be noted that the present results could also be taken to indicate that certain aspects. . . . Overall, the present results have broad theoretical implications in that they show that language specific processes interact with visual processes . . .

Thus, the topical interests of the great high medieval scholars in Western Europe were subsequently abandoned in the Enlightenment, and along with them the traditional overt form of the recursive argument method, but it is still used, though less frequently and in altered form, as the recursive scientific method.

The old recursive argument method actually continues to live on in the "humanities" too, though less formally. It constitutes the typical argument structure of an academic dissertation, which is the medieval *dissertatio* in origin, the document an advanced university graduate student presents in order to be awarded the degree of Doctor of Philosophy, or Ph.D.[37] This deformalized recursive argument method is also found quite frequently in research monographs, many of which are derived from academic dissertations.

In the absence of the characteristic overt structure of a recursive argument method work, with its lists of subarguments, a typical dissertation is hardly distinguishable very clearly from a treatise. Nevertheless, a traditional dissertation is often characterized by an expressly stated TOPIC; an INTRODUCTION in which arguments for and against previous work on the topic, and often specific hypotheses as well, are stated; the AUTHOR'S EXPOSITION; and a CONCLUSION, in which the original arguments for and against the topic, and the specific hypotheses, if any, are reevaluated in the light of the results of the data analysis and arguments presented by the author in the main thesis section. The structure of an "ideal" humanities dissertation may be analyzed as follows:

### Ideal Structure of a Modern Humanities Dissertation

I. [MAIN ARGUMENT]

[topic, question]

[37] This same degree, Ph.D., is awarded to students in the sciences, the social sciences, and the humanities. It is typically only the professions that award other doctoral degrees, e.g., M.D., Doctor of Medicine.

II. Introduction

[previous views, arguments for and against previous views, presentation of hypotheses]

III. [Author's View (*Exposition, Body of the work*)]

[author's own views]

IV. Conclusion

[arguments against arguments in II, based on results in III, evaluation of hypotheses]

The "ideal" dissertation therefore has the structure of a single large-scale recursive argument. This kind of structure appears to derive directly from the Medieval Latin recursive argument method, as the dissertation itself derives from the medieval *dissertatio*. However, in comparison with such a medieval work, the modern dissertation is distinguished by its looseness of structure, its carefree (and often careless) logic, and the fact that authors rarely respond once again to all the statements given in the Introduction for and against previous views. Moreover, despite its approximate formal resemblance to a recursive argument, the actuality of a modern dissertation is typically devoid of the scientific approach that characterizes medieval texts written according to the traditional recursive method.

As pointed out in chapter 2, the recursive argument method can theoretically be used to investigate any topic at all. Some scholars have rightly noted that the method encourages logical investigation of problems and opens the door to possibilities that might not have been imagined without the demands of the method, with its exhaustive lists of arguments and counterarguments, author's arguments, and refutations. Although the dedication of premodern scholars to precise, logical thinking—as demonstrated by the great medieval and Renaissance texts written according to the recursive argument method—has to some extent become alien to scholarship in the modern world, perhaps one can hope for a revival someday.

# Appendix A
# On the Latin Translations of Avicenna's Works

The earliest example in Latin of the recursive argument method or *quaestiones disputatae* format is found in Avicenna's *De anima* 'On the Soul'. Although translated and circulated as an independent work, it is actually only a part of Avicenna's huge *summa*, the *Kitāb al-shifā'*,[1] 'The Book of the Healing', from which the *Metaphysics* was also extracted and translated, most likely at around the same time, under the sponsorship of the same patron.[2] *De anima* was translated[3] into Latin by *Avendauth israelita philosophus* 'Ibn Dā'ūd the Jewish philosopher', aided by *Dominicus archdiaconis* 'Dominicus the archdeacon',[4] and dedicated to Archbishop

[1] The Arabic name is given in Latin as *Asshiphe* (d'Alverny 1961: 289), a good transcription of the Arabic as pronounced, *Ashshifā'*.

[2] Van Riet (1972: 99* n. 26) remarks that the Latin of passages of the *De anima* incorporated into the *Summa theoricae philosophiae*, a work attributed to Algazel (al-Ghazālī), are remarkably different from the Latin of the same passages in the *De anima* translation itself, suggesting that these two particular works were not done by the same *Dominicus archdiaconis*. She also notes that Algazel's work was translated in Toledo in the second half of the twelfth century and is attributed to Gundissalinus, but "la terminologie utilisée ne recoupe pas celle que l'on peut lire, pour des termes arabes identiques, dans la traduction latine de la *Métaphysique* d'Avicenne" (Van Riet 1977: 135*–136*). Finally, Van Riet (1972: 123*) remarks that the Latin translation of the *De anima* does not contain a single transcribed Arabic word, unlike the translations of the *Canon* (on medicine), the *De animalibus*, and the *Physics*. She does not mention whether or not the *Metaphysics*, like the *De anima*, is distinguished in this way, but an Arabic-based usage shared by Gerard of Cremona (d'Alverny 1952: 337–366) suggests that the *Metaphysics* is not an exception. The lack of examples in the *De anima* would seem to be due to the deep knowledge of Avendauth, who was presumably better able than nonphilosophers to explain everything to his young Latinist collaborator.

[3] Van Riet (1972: 98*) cites solid textual evidence that the translation was made not orally but in writing. It is worth noting that just as the *Avicenna Latinus* editor discovered that there are two different versions of the text of the *De anima*, so too are there two different versions of the text of the *Metaphysics* (Van Riet 1977: 128*).

[4] Avicenna (Van Riet 1972: 3–4). He is not further identified in any manuscript of this particular work. Van Riet (1972: 98*–99*) remarks that there is no reason to identify him with the archdeacon Domingo Gundisalvo (or Dominicus Gundisalvi, Gundisalvus, or Gundissalinus), to whom translations and new philosophical works written in the following decades are attributed. However, on the basis of the dated documentary evidence it seems clear that from March 11, 1162, to December 1, 1178, one and the same person—who in legal documents is regularly called *Dominicus Colarensis archdiaconis* 'archdeacon of Cuellar' (in the diocese of Segovia, which was then united with Toledo)—is intended (d'Alverny 1989: 196 n. 4). There seems no reason to doubt that *Dominicus archdiaconis*, Avendauth's helper for his translation of the *De anima* of Avicenna, was this same

John of Toledo (r. 1152–1166),[5] though the project was probably begun under his predecessor, Archbishop Raimund (r. 1125–1152), who sponsored at least one translation project.[6] Avendauth is undoubtedly to be identified with Ibrāhīm ibn Dā'ūd (i.e., Abraham son of David), the well-known Jewish philosopher who lived in Toledo from ca. 1148 until his death in 1180.[7] His Latinist assistant, Dominicus the Archdeacon, has been identified with Dominicus Gundisalvi, archdeacon of Segovia and Toledo.[8]

Dominicus and the prolific translator Gerard of Cremona were clerical colleagues in the episcopate of Toledo and may have worked together on one or more translations. At first Dominicus worked with Avendauth as his Latinist assistant; later he worked with an Arabist assistant known as Master John.[9] In addition to translating some of the works of Avicenna, they translated a number of other important works, including

---

person. He would then have been young and inexperienced, and very much under the influence of the great Jewish philosopher who headed the translation project, from whom he certainly learned much, as his own works are full of quotations of Avendauth.

[5] Van Riet (1972: 95*).

[6] d'Alverny (1952: 356; 1982: 446); cf. Fidora (2003: 98 n. 6). The project sponsored by Raimund was a translation of a philosophical work in Arabic by the Christian writer Quṣṭā ibn Lūqā (Fidora 2003: 105, 196).

[7] d'Alverny (1954; 1982: 445–446, 451); cf. Burnett (1998). This identification has been much doubted, for dubious reasons. The fact that Avendauth was earlier confused by scholars with the later collaborator of Dominic named *magister Iohannes* 'Master John' (d'Alverny 1982: 445) hardly means that the identification of Avendauth need be doubted. *Avendauth* is the precise, regular Andalusian transcription of the Classical Arabic name Ibn Dā'ūd. It is possible that there were two Jewish scholars with exactly this same name who lived at exactly the same time in Toledo, but that they both were *philosophers* and were well acquainted with the Neoplatonic Aristotelianism of the Classical Arabic tradition would seem to make it as certain as anything can be in medieval studies that there was only one of them, namely Ibrāhīm ibn Dā'ūd, or 'Abraham son of David'. It is he who brought "the importance of the *al-Shifa' (Healing)* of Avicenna (Ibn Sina) . . . to the notice of Archbishop John of Toledo" (Burnett 1998). Cf. Van Riet (1972: 101* n. 32).

[8] The theory of Rucquoi (1999) that there were two men with similar names has mostly been rejected by other scholars, e.g., Hasse (2004: 73 n. 30); see also Fidora (2004). His name is spelled in manuscripts Gundisalvi, Gundisalvus, Gundissalinus, etc.

[9] d'Alverny (1982: 451). However, as she notes in her earlier study (d'Alverny 1952), the individual sections of the *Kitāb al-Shifā'* (of which the *Metaphysics* is only one part) are attributed variously to Avendauth alone, Avendauth and Dominicus (Gundissalinus), or to (Dominicus) Gundissalinus alone, while some are anonymous (d'Alverny 1982: 446 n. 104). Burnett (1998) adds, "Portions of the text, including those on universals, physics (in part), the soul and metaphysics, were translated by Avendauth, Dominicus Gundissalinus (an archdeacon in the cathedral, *fl.* 1161–81) and a certain '*magister* John of Spain'." The *Metaphysics* translation is also attributed to Gerard of Cremona (d. 1187), who was working in Toledo at the same time and is named as translator on one of the earliest manuscripts (Van Riet 1977: 123); Gerard is also named together with Dominicus on a document from the Toledo archives dated 1176 (d'Alverny 1982: 543–454). Nevertheless, it is believed that Gerard did not actually work on the *Metaphysics*.

the *Maqāṣid al-falāsifa* 'The Aims of the Philosophers', an Arabic version by Algazel (al-Ghazālī) of Avicenna's short summary of Islamic Aristotelian philosophy.[10] Dominicus also authored several works of his own, all of them strongly influenced by Avicenna and the other Classical Arabic writers whose works he and the others working in Toledo at that time translated. He authored several influential philosophical works in Latin, including a work entitled *De anima*, which consists in great part of verbatim quotations of Avicenna's *De anima*. His works show that he had a good knowledge of philosophy and of theological works written in France in his own time, indicating that he may have studied there.[11] This possibility is supported by his evident lack of much personal knowledge of Arabic. Although d'Alverny's other suggestion—that French clerics could have brought books from France to Toledo—would seem to be an unlikely example of "carrying coals to Newcastle" for that period, in fact there was a great deal of coming and going, which accounts for the increasingly rapid diffusion in France both of the translations and of new works such as those by Dominicus, which were evidently transmitted to France in the late twelfth century, shortly after they were completed, along with other translated works by Avicenna, Algazel, Aristotle, and many others. Dominicus's own works were quite influential and were still cited in the thirteenth century by the great theologians of the day.

The dates of Dominicus have been much discussed. Noting that he is mentioned in documents as late as 1181, d'Alverny says that therefore his *floruit* must be shifted accordingly,[12] but this by no means follows. He may well have lived a very long time, but what is significant to history is his activity as a translator and author, in which capacity he is firmly attested during the time of Archbishop John of Toledo (1152–1166); his *floruit* as a scholar is thus properly 1152–1166. His name is attested on legal documents from 1164 to 1181,[13] so that period is his *floruit* as a church official. Putting the two together, his known career ran from approximately 1152 to 1181. The center of his *floruit* as a whole is thus 1167–1168, only one or two years after the death of his patron, Archbishop John. This suggests

[10] Burnett (1998). Avicenna's work is now known in its Persian form as the *Dāneshnāmeh*, q.v. chapter 7, note 58.

[11] d'Alverny (1989: 197). Cf. especially Jolivet 1988: 138–139), who refers to Hugh of St. Victor and to a specifically "Chartrian tradition" (q.v. Fidora 2003: 19). Jolivet's article gives a good brief overview of the works authored by Dominicus. For detailed examination of his epistemology, see Fidora (2003, 2004).

[12] d'Alverny (1982: 445).

[13] d'Alverny (1989: 196 nn. 4–5). A "Dominicus Gundisalvi" also occurs on a document from Palencia, possibly referring to the same person (Fidora 2003: 15).

that most of his translations were done (or at least begun) under John's patronage.

It is unknown precisely when Avicenna's own works became available in Paris, but it is quite clear that his thought was known there, directly or indirectly, by no later than the end of the century. Avicenna's works, which were at first thought by Medieval Latin scholars to be by Aristotle,[14] and Avicennist thought such as that incorporated into the *De anima* of Dominicus Gundisalvi and the anonymous *Liber de causis primis et secundis*, another early Avicennian work,[15] had such a significant effect on theology and metaphysics in Paris that in 1210 (and again in 1215, and yet again in 1225) the Church instituted formal restrictions on the reading of many of the works of Aristotle and his unnamed commentators.[16] At that time this apparently applied mostly to the translations into Latin of Arabic works, the original authors of which were nearly all Muslims. The direct motivation for the restrictions is thought to have been the unorthodox views on the Trinity, on the divinity and humanity of Jesus, and on other issues, some of which were considered Avicennian in origin.[17] The restrictions seem to have had virtually no effect, as shown by the necessity of repeatedly promulgating them.

At any rate, the known period of Archbishop John's tenure in Toledo dates the translation of Avicenna's *De anima* securely to 1152–1166, thus also dating the earliest identified example[18] of the recursive argument method in Western Europe. The next text to contain examples of the method, the Latin translation of Avicenna's *Metaphysics*, should probably be dated to around the same period.[19]

[14] d'Alverny (1982: 451).

[15] Whatever label one wishes to put on it, it cannot be denied that it depends heavily on Avicenna, as shown by Vicaire (1937)—a work that has not yet been replaced—and by the edition of Vaux (1934). The latter's clarification of what he means by calling the work "Avicennist" (Vaux 1934: 134) seems to have been ignored by subsequent scholars.

[16] Vicaire (1937: 450–451); Vaux (1934: 70), who says of the *Liber de causis primis et secundis*, "Plus généralement, la synthèse qu'il tente d'Avicenne et d'Augustin, du néoplatonisme arabe du *De causis* et du néoplatonisme chrétien de Denys représenté par Érigène, évoque l'atmosphère mélangée des décrets de 1210–1215, où furent atteints à la fois Aristote et ses disciples arabes, et au moins indirectement Érigène."

[17] Vicaire (1937).

[18] Example 6.1, given in chapter 6, is simply the first one I found in the Latin text that was obvious to me.

[19] See note 9 in this appendix.

# Appendix B
# On Peter of Poitiers

According to the traditional chronology of the partly edited text of the *Sententiarum libri quinque* 'Five Books of Sentences' of Peter of Poitiers (d. 1205),[1] the text as we have it is the same as the text that was dedicated to William of Champagne by 1176. It this were true, the work would be the earliest known native Medieval Latin composition that uses the recursive argument method,[2] since the otherwise earliest identified text to use the method is the *Summa* of Robert of Curzon, which has traditionally been dated to 1202.[3] However, upon investigation of the text of the *Sententiarum libri quinque* (henceforth, *Sentences*), its author, and the historical circumstances, a number of insurmountable problems are found that not only rule it out as the earliest datable work to use the method but also raise questions regarding the dating and reworking of the text over time, its modern edition, and the identity and floruit of Peter of Poitiers himself.

Traditionally, the *Sentences* of Peter of Poitiers (b. in Poitiers ca. 1130, d. 1205 in Paris), his most important work, is dated to between 1170 and 1176.[4] It has generally been believed that he largely follows the views of Peter Lombard, on whose *Four Books of Sentences* he based his own *Sentences*, so he is usually thought to have been the Lombard's student. However, as shown conclusively by the editors of the first two books of Peter of Poitiers' own magnum opus, none of his references to *magister meus* 'my teacher' can refer to Peter Lombard.[5] This is significant in connection with the date of the *Sentences* of Peter of Poitiers.

Moore (1936: 25ff.) describes the manuscripts, date, and historical context of Peter of Poitiers' *Sentences*. It is unfortunate that, as he and

[1] Given as *Sententiae Petri Pectaviensis* 'The Sentences of Peter of Poitiers' in the partial edition by Moore, Dulong, and Garvin (1943–1950).

[2] It also would be a practically unprecedented work for the period, when texts like Peter's *Sentences* were regularly rewritten, expanded, contracted, changed in doctrinal or theoretical orientation, and just about everything else imaginable.

[3] See below on the recently proposed lower dates.

[4] Moore and Dulong (1943: vi).

[5] Moore and Dulong (1943: xlvi–l). It appears that no one has followed up on Moore and Dulong's careful analysis.

Dulong (1943–1950) note, their edition of the first two books is based more or less exclusively on four manuscripts (E, B, Lr, Ts), which represent only three out of their seven manuscript families (including a total of twenty-four complete manuscripts) and do not include any of the twelfth-century Paris manuscripts. It is also unfortunate that the edition of this important work was never completed. A full edition is a great desideratum for medieval studies.

Not a few scholars (including Moore 1936: 41)[6] have commented that it is hardly conceivable that Peter could have written the work in 1170 if he became master of theology in Paris only in 1169, as is generally believed. The work is now thought to have been completed by 1175–1176, based on the dedication in the work's prologue to William of Champagne, archbishop of Sens (1168–1176), who was appointed "archbishop of Rheims, where he was received on August 8," 1176 (Moore 1936: 27, 39–40). Peter's dedication to the archbishop is paraphrased in the *Contra quatuor labyrinthos Franciae* 'Against the Four Labyrinths of France' (Glorieux 1952: 304) by Walter of St. Victor,[7] whose work is dated to 1177–1178 by Glorieux (1952: 194–195) on speculative and doubtful grounds;[8] there is in fact no certain terminus post quem for the latter work other than the death of Walter in 1194.

There has long been confusion over the different personages of the twelfth and thirteenth centuries known as Peter of Poitiers, and it appears that despite much discussion in the mid-twentieth century some questions still exist. In addition to the supposedly unitary Peter of Poitiers who wrote the *Sentences*, there are two other men, from approximately the same period, who are also usually known as Peter of Poitiers, namely: (1) a canon of Saint Victor, author of a *Penitential* and some minor works, who died after 1216 (Moore 1936: 23–24); and (2) a monk and prior of Cluny who was a poet and secretary of Peter the Venerable, whom he accompanied to Spain in 1142–1143.[9] He participated in the translation of the Koran and other texts from Arabic into Latin there,[10] and died in 1160

[6] However, Moore (1936: 7–8) also says that the *Sentences* "at the latest appeared near the end of 1175." Nielsen (1982: 281, 311), despite noting that the 1170 dating "is somewhat problematical," follows everyone else in using it as the linchpin of practically the entire chronology of late twelfth-century thought (e.g., Nielsen 1982: 342–344 n. 227), the thoroughgoing weakness of which begs for a determined, informed deconstruction.

[7] He was prior of the monastery of St. Victor from 1173 to 1194.

[8] This seems to have been widely overlooked.

[9] Kritzeck (1964: 10–14). Actually, there are still other men known as Peter of Poitiers in approximately the same period, but they are typically distinguished clearly in recent scholarly literature and therefore are not discussed in this already lengthy appendix.

[10] Kritzeck (1964: 56, 60–61).

(Moore 1936: 21). It is probable that some of the confusion involves one or both of the other Peters of Poitiers, but there are many more pressing questions that need attention, so for the most part the following discussion is limited to the problem of reconciling some apparently contradictory facts concerning the Peter of Poitiers who is generally believed to have been master of theology, Chancellor of Paris (i.e., of the cathedral school of Notre Dame), and author of the *Sententiarum libri quinque* (among other works). The abbreviation PP is henceforth used to refer specifically and exclusively to this particular Peter of Poitiers.

According to Alberic of Trois-Fontaines, PP taught theology in Paris for thirty-eight years. If so, this would mean he became master in 1167 and taught until his death in 1205 (Moore 1936: 6), including a period of two years before he succeeded Peter Comestor in the chair of theology in 1169, and for twelve years after he became Chancellor of Paris in 1193 (Moore 1936: 6). There is something seriously wrong with this dating. Moore (1936: 6) notes that when a master of theology was appointed to the position of Chancellor of Paris (also referred to as "Chancellor of Notre Dame"), he gave up his chair in theology: "It is very doubtful . . . that Peter of Poitiers continued to teach after his appointment to the chancellorship of Paris in 1193. Stephen D'Irsay says that the chancellor supervised the classes but did not share in the teaching. We know, furthermore, that Peter Comestor gave up his chair of theology in 1169, shortly after being named chancellor. We may suppose, therefore, that Peter of Poitiers quit the classroom in 1193. This agrees with the statement of Odo of Cheriton (1247) that our author taught theology 'more than twenty-four years.'" The latter option, according to Moore, means the twenty-four years between 1169 and 1193, the date of PP's succession to the chair of theology of Peter Comestor. However, as a conclusion to his discussion of the date of PP's *Sentences*, Moore (1936: 41) declares, "Peter of Poitiers wrote his *Sentences* at Paris. Of this there can be no reasonable doubt. Alberic of Trois-Fontaines informs us that at the date of his death (1205), Peter of Poitiers had taught theology there for thirty-eight years, or since 1167.[11] Hence it was during the first years of his theological teaching at Paris, that he wrote this work." In view of the cited data, Moore's account is bewildering.

PP's most famous work is the *Sentences*, which is devoted to disputable problems in theology and is characterized throughout by logical

[11] Moore (1936: 41 n. 31) here quotes Alberic: "*Obiit magister Petrus Pictaviensis, cancellarius Parisiensis, qui per annos 38 theologiam legerat Parisius.*" However, he himself casts doubt on Alberic's dates (Moore 1936: 6).

dialectic. The work has been characterized by most scholars, on the basis of its dominant orthodox Augustinism, as a work of the school of Peter Lombard, on whose own famous *Sentences* PP heavily depends. However, it is now known that the book was influenced by the work of Simon of Tournai, a leading member of the Porretans, students or followers of Gilbert de la Porrée (Gilbert of Poitiers). It has also been explicitly demonstrated that PP's teacher, whom he calls *magister meus* 'my master', was *not* Peter Lombard (Moore and Dulong 1943: xlvi–l) and, in fact, PP differs from the Lombard more often than not (Moore et al. 1950: xxix–xxx).[12]

All of the other works attributed to PP apparently belong to the period after he became Chancellor of Paris in 1193 (Moore 1936: 6), and are different from the *Sentences* in content and approach, so they cannot be used to help date it. According to Moore (1936: 58–59), the *Allegoriae super tabernaculum Moysis* is attributed in the early manuscripts "to a *magister* Peter of Poitiers, and this term *magister* would have been used in referring to the chancellor rather than to the religious of Cluny or to the canon regular of Saint Victor's." The manuscripts of the *Distinctiones super psalterium* attributed to Peter of Poitiers regularly call him a *magister*.[13] The *Distinctiones super psalterium* is undated, but Moore (1936: 91) says it probably should be dated "not many years before the *terminus ad quem* date, 1196." However, titles and authors given in manuscripts are typically much later in date than either the work itself or the putative author's dates; they are well known to be far from certain accounts of either titles or authors. Again, PP—according to Moore "the chancellor, as theologian"—is the author of the *Compendium historiae in genealogia Christi* (Moore 1936: 108). He also argues that a "*master* Peter of Poitiers" (Moore 1936: 121, italics his) was the author of the *Historia actuum apostolorum*, "which forms the last part of the *Historia scholastica*" dated to shortly before 1183 by Moore (1936: 119–122). The latter work was formerly attributed in its

[12] "We have said that Peter of Poitiers introduces many questions into his *Sentences* which are not in the work of Peter Lombard, and that even when common questions are under discussion his treatment of them is largely different. It is not astonishing, therefore, that his text is for the most part personal and independent of the Lombard text" (Moore and Dulong 1943: xxix–xxx).

[13] One thirteenth-century manuscript uniquely gives the title of the work as *Distinctiones super psalterium edite a magistro Petro Pectaviensi, cancellario iuris* (Moore 1936: 85). The author is thus called both a "master [of theology]" and a "chancellor of [canon] law." This is striking, if not incredible, because the author of the *Sentences* does not discuss legal questions at all, suggesting that he did not know much about it (cf. below). However, this could be the kind of random error or peculiarity typical of manuscript attributions (*iuris* instead of *paris[iensis]* or the like), and might not be historical.

entirety to Peter Comestor,[14] but there is still some doubt about whether one of the other Peters of Poitiers might be the author. Finally, all of PP's preserved *Sermones*—of which there are "at least 59" according to Moore (1936: 171)—are undated, but Moore (1936: 137 et seq.) says, "some, if not all, of these sermons belong to the late years of our author, and very probable [*sic*] to the time of his chancellorship (1193–1205)." It would seem desirable to reexamine some of the other works attributed to PP by Moore to confirm them or possibly reassign their authorship to others.

PP's obituary notice in the records of the cathedral of Notre Dame says that he died on September 3,[15] and calls him "*magister* 'master [of theology]' Petrus Pictaviensis, *diaconus et cancellarius* 'deacon and chancellor'."[16] Other documentary evidence on PP includes an act of the bishop of Paris, Maurice, dated to 1184, which is attested to by, among others, a *Magistro Petro Pectavensi* (Moore 1936: 10–11). PP is mentioned in an act of 1193, in which he is explicitly called a "chancellor" (Moore 1936: 11–12). Moore notes the existence of "seven acts of the chancery of Notre Dame given by the hand of Peter of Poitiers" dating to between 1193 and 1204. The documentary evidence for PP's dates include four papal letters to him, all of them strictly legal in nature, dated January 14, 1196, November 4, 1202, ca. 1204, and April 1, 1205 (Moore 1936: 16–17).

The last point would seem to be highly significant. Moore (1936: 5) says, "Peter of Poitiers most probably studied law. In any case, he had a practical knowledge of this subject, especially of procedure, that was far from rudimentary, for . . . the Holy See several times named him judge-delegate in ecclesiastical trials.[17] In his *Sentences*, however, PP scrupulously avoids questions belonging to the canonists rather than to the theologians. In this he is an exception to the rule, for in most works of the period is found a mixture of canon law and theology." But if PP was an expert on law why does he "scrupulously" never discuss matters of law and judicial procedure in his *Sentences*? This is extremely peculiar. One should think rather that the author did not discuss law because he

[14] Peter Comestor dedicated his work to William of Champaigne, archbishop of Sens (1168–1176), the same person to whom PP dedicated his *Sentences* (Moore 1936: 118–119). Peter Comestor retired from teaching to St. Victor in 1169.

[15] The year is missing and has been supplied from the histories (which are also defective); see Moore (1936: 17ff.), who gives 1205, as discussed above.

[16] This is in the *Cartulaire de l'Eglise de Notre Dame de Paris*. It reads in full: "III Non. Septembris. De domo sancte Marie obiit magister Petrus Pectaviensis, diaconus et cancellarius, qui dedit nobis quadraginta libras Parisiensium, positas in emptione decimi Viriaco" (Moore 1936: 17).

[17] Moore (1936: 17) notes that the papal "letters indicate that Peter of Poitiers had a practical knowledge of law and judicial procedure."

did not know anything about it. Yet PP was chosen by the pope himself on several occasions as his legal representative in matters of canon law. Moreover, the orthodoxy of PP's *Sentences* was never called into question by the authorities, unlike the works of the others criticized by Walter of Saint Victor. Was PP the chancellor actually a canon lawyer and different from PP the theologian? If not, why did the pope not appoint a canonist? Furthermore, Moore (1936: 6–7) states, "The works of Peter of Poitiers show that he was interested in three branches of theological study: theology properly so called, or the dogmatic and moral questions to which the study of Scripture and the Fathers of the Church gave rise; sacred history; and the spiritual or allegorical interpretation of Holy Writ which constituted medieval exegesis. His *Sententiarum libri V*, a systematic, comprehensive work on dogmatic and moral questions, is the product of his teaching in theology proper." Moore emphasizes throughout his discussion of PP's *Sentences* that PP was interested almost exclusively in disputabilia, so he omitted discussion of other topics. But the most disputable topics of all were legal, as PP would have known if he were trained in canon law. Why then did he fail to discuss legal questions, which are discussed by more or less all other theologians of the time, including Robert of Curzon in his *On Usury* (an extract from his *Summa*), which contains examples of the classic recursive argument? This would be likely if the PP who authored the *Sentences* was not, in fact, educated in law but in theology, whereas the PP who was chancellor was educated in law. However, scholars seem not to have contested the assumption that there was only one PP, who was in turn master of theology and Chancellor of Paris, like Peter Comestor and others before and after PP.

Moore notes that PP discusses the meaning of Greek terms (Moore 1936: 4–5 and infra), though he thinks PP may have simply copied them from works of his predecessors. In fact, although it is unlikely that PP knew Greek, his handling of Greek citations indicates that he had at least studied with someone who had some actual knowledge of the language. This suggests the school of Gilbert de la Porrée (Gilbert of Poitiers, d. 1154), who "had a special reverence for the Greek Fathers; he and his followers were active agents in the revival of Eastern patristics."[18]

As Chancellor of Notre Dame, PP was also a close associate of Peter the Chanter (d. 1197), who held the position of *Cantor* 'Chanter' there, and "on at least one occasion they served together as papal judge delegates" (Baldwin 1970: 44). In theological doctrine, the Chanter "was prone to

[18] d'Alverny (1991: 431); see Häring (1962) for a detailed study.

adopt the solutions of the Porretani, particularly those of Alain of Lille" (Baldwin 1970: 48).

## The Saint Victor Connection

As Moore and Dulong (1943: xlvi–l) show, PP's unidentified teacher, whom he calls *magister meus* ('my teacher'; he also refers to *magistri nostris*), was definitely *not* Peter Lombard,[19] to whom he refers explicitly twice as *magister in libro sententiarum* and *in libro sententiarum Petri.* Moreover, as noted above, despite his citation of many passages from Peter Lombard's *Sentences*, PP very often disagrees with him. Who then was his teacher, and where and when did he study? The fact that PP shared some of Peter Lombard's views on the Trinity—the topic of the first of PP's *Five Books of Sentences*, referred to in the prologue of the work as *De fide Trinitatis* (Moore 1936: 41)—is significant, but the views for which PP, Peter Abelard, Peter Lombard, and Gilbert de la Porrée are castigated by Walter of Saint Victor[20] all center on the Trinity. One notes in this connection the possible influence of a number of important scholars of the time, including Gilbert de la Porrée, a master whose doctrines on the Trinity were condemned by the Church but who remained respected nevertheless, especially for his *Commentary on Boethius'* De Trinitate,[21] and Richard of St. Victor (d. 1173), whose most famous strictly theological work is *De Trinitate.*

In addition, it is now accepted that chapter 22 of Book I of PP's *Sentences* "is identical with sections 4–10 of the introductory portion of Simon of Tournai's treatise on the Trinity, contained in his *Institutiones in sacra pagina* (or *Summa*)," and that "sections 1–10 of Simon's *Institutiones*, the only part in which passages identical in that work and Peter of Poitiers' *Sentences* have been found, occur word for word in the *Expositio Symboli S. Athanasii* (or *Tractatus super Quicumque vult*) ascribed to Simon. . . . It would seem, therefore, that Peter used the *Expositio*" (Moore et al. 1950: xxxvi–xl). Simon of Tournai was one of the leading Porretans, and it would be odd for someone opposed to that school's positions to support it in this way, so textually, this chapter is an interpolation into PP's work.[22] Yet the very fact of an interpolation of an entire

[19] For the *Sentences* of Peter Lombard, see Silano (2007).

[20] Glorieux (1952: 194).

[21] See Jacobi (1998) and Marenbon (2002) for discussion of his philosophical views.

[22] See below on the chronological significance of the interpolation.

chapter opens the door to several other potentially crucial problems with the text of PP's *Sentences*.

It must also not be forgotten that one of the other contemporaneous Peters of Poitiers was a canon of Saint Victor who lived in Paris at the same time and authored a number of works. He was a member of the circle of Peter the Chanter, PP's colleague at Notre Dame (Baldwin 1970: 33–34). Although Peter of Poitiers "of St. Victor" is generally considered to have had a low position compared to others in the Chanter's circle, he is explicitly named as a master on at least one legal charter he witnessed in 1209. He was also wealthy and had much to do with other members of the circle.[23] He was not, therefore, secluded in the cloister of St. Victor, no more than were two other famous canons from the same institution, Hugh of St. Victor in the early twelfth century and Walter of St. Victor in the late twelfth century, whose fame derives from their scholarly writings. In his *Penitentiale* he cites a wide range of legal authorities, from papal decretals to opinions of individual legal scholars.[24] He thus had practical and theoretical knowledge of canon law exactly appropriate to the functions performed by PP for the pope. It seems quite likely that Peter of Poitiers of Saint Victor has been much too hastily dismissed.

## The William of Champagne Connection

As noted above, the prologue of PP's work dedicates it to William "Whitehands" of Champagne, archbishop of Sens from 1168 to 1176 and uncle of King Philip II Augustus of France. William was appointed archbishop of Rheims, where he was received on August 8, 1176 (Moore 1936: 39). Moore says this dedication is not otherwise significant, since other writers of the time also dedicated works to that very powerful man. But the conclusion is drawn by Moore (1936: 40): "Since he still addresses William as archbishop of Sens, his work must have been completed before the beginning of the year 1176." Furthermore, Pope "Alexander III definitively condemned the Lombard's teaching that *Christus secundum quod est homo, non est aliquid* in a letter to William of Champagne, become archbishop of Rheims, dated February 18, 1177. . . . But the question was still heatedly discussed in the Third Council of the Lateran in 1179" (Moore 1936: 40 n. 30). If PP's book was already written and known (e.g., by Gauthier of Saint Victor), why did the pope not mention PP? Why did the Lateran Council

[23] Baldwin (1970: 33).
[24] Baldwin (1970: 33–34).

not discuss PP's book, or demand that he attend the meeting to defend the Lombard's position? This is incomprehensible. Moreover, neither the pope's letter nor the Lateran Council of 1179 had any effect on PP's book. If the book had been published in 1175–1176, as hitherto supposed, this is inconceivable. The implied belief that the dedication to William and its citation in the *Four Labyrinths* mean PP must have finished the work as we know it by that time, and that it could not have been substantially changed afterward, is disproven by the interpolation of an entire chapter—and who knows what else—into PP's apparently still unedited *Sentences.*

## The Recursive Method and the *Sentences* of Peter of Poitiers

It is also necessary to consider the surprising precociousness of PP's use of the recursive argument method in the received text of his *Sentences.*

There exists a shorter version of PP's *Sentences*, preserved in at least four manuscripts dating to the fourteenth century (Moore and Dulong 1943: lx). Moore (1936: 174–177) gives several excerpts from this shorter version, which he says is "abbreviated." None of the five arguments he quotes from it use the recursive argument method. Because the manuscripts date from a period when the method was virtually universally understood and very widely used, this is puzzling (assuming that the corresponding texts in the "long" version do, by contrast, use the recursive method). It is therefore possible that the shorter version of the *Sentences* descends from the original, which was presumably sent to William. Similarly, the quotations of PP's text in Walter of St. Victor's *Four Labyrinths of France* do not seem to use the recursive method.[25] The recursive method is otherwise unattested in "native" Latin works until the *Summa* of Robert of Curzon (ca. 1202, recently redated to ca. 1208–1215), about twenty-five years after the date derived from the prologue of PP's *Sentences.*[26] The manuscripts of the shorter version of the *Sentences*, and the putative twelfth-century manuscripts of the "long" version, neither of which were used in Moore et al.'s edition of the first two books of PP's *Sentences*, should be examined carefully to see whether they shed any light on the development of the text over the decades of PP's career. Valente, in response to similar concerns, suggests that the uncertainty about the relative chronology of Peter of Poitiers' *Sentences*, vis-à-vis the works of

[25] The quotations of PP and the text of PP's *Sentences* in Walter's *Four Labyrinths* need to be examined closely by a specialist.

[26] Macy (2009). Macy's date is uncomfortably close to that of Robert's death in the Holy Land.

Simon of Tournai and other near contemporaries, indicates "the probability that the redaction of his works took place over a period of some years."[27] This scenario is supported by some additional odd usages and chronological peculiarities in PP's work that are noted by her.[28]

Moreover, it is pointed out by Baldwin (1970: 24–25, 100–101) that Robert of Curzon's *Summa* (ca. 1202) consists in large part of a reorganization and modernization of the *Summa* of Peter the Chanter, as Geoffey of Poitiers' *Summa* is largely a reorganization and modernization of Stephen Langton's *Questiones*. "Peter the Chanter's *Summa de sacramentis* was a collection of *questiones* never fully organized or completed. Using one version of the work [preserved in manuscript W], Robert of Courson revised the general organizational plan to emphasize penance and morality, and within each *questio* reconstructed the material according to a methodical procedure. To be sure, Robert was the author of a new *Summa* but in another sense his book constituted another draft of work begun by the Chanter. In the last analysis, the *Summa* of Robert of Courson was a final product of the school of Peter the Chanter."[29]

It appears that PP's work underwent much the same process. PP evidently wrote a version of his *Sentences* soon after he began teaching in Paris, and dedicated it to William of Champagne—perhaps due to a suggestion by Peter Comestor, who had done the same thing. This original *Sententiarum libri quinque* (preserved in manuscripts as the "abbreviated" version of the text?) would have used the old-fashioned *quaestiones* method. Subsequently, PP or one of his students reworked the early version of his *Sentences*, probably over a considerable period of time, during which the verbatim quotation of Simon of Tournai's text was interpolated and some of PP's arguments were rewritten according to the new recursive method, as Robert of Curzon did when he updated Peter the Chanter's work as his own *Summa*. The original dedication of PP's work could hardly have been deleted, even if William (d. 1202) was dead by then: he was not only a powerful member of the ruling elite and uncle of

[27] Valente (2007: 28).

[28] She remarks, "la *Summa* d'Alain [de Lille] présente un système logico-sémantique relativement systématique et cohérent . . . , tandis que quelque chose de semblable manque complètement chez Pierre de Poitiers" (Valente 2007: 387 n. 387), and significantly, "Ni son recours aux *fallaciae* ni son usage de la notion de *suppositio* ne montrent, en conclusion, une assimilation approfondie, de la part de Pierre de Poitiers, de la réflexion logique de son époque. . . . Il utilise abondamment des notions d'origine porrétaine—l'influence de Simon de Tournai sur Pierre de Poitiers est démonstrée, et l'influence d'Alain de Lille sur lui est aussi probable—, mais ces imprunts aux porrétains sont coupés de leur contexte epistémologique d'origine et de la théorie sémantique organique dans laquelle elles avaient vu le jour" (Valente 2007: 333).

[29] Baldwin (1970: 25).

the king, Philip II Augustus, he had personally crowned him in 1179, and Philip ruled until 1223.

This scenario would explain much that has so far seemed inexplicable in PP's *Sentences*, including the relative conservativeness of its general approach combined incongruently with some Porretan elements and the use of the Logica Nova, as well as the hitherto overlooked precociousness of its use of the recursive argument method. In addition, it must be noted that in the received text of the *Sentences*, PP very clearly implies that the recursive argument method he uses differs slightly from what must have been the norm, on which he comments. In the Author's View Argument section in Example B.1 below, PP remarks, a little oddly, "It does not matter much whether it is said that the willing and the act are the same sin or different [sins]. We respond therefore both to those which are objected against the first part as well as those against the other [part]." This means that he provides Subarguments$_2$ in reply to all the Subarguments$_1$— suggesting that the normal practice, which must therefore already have been well established (suggesting a considerable period of development), was for an author to omit replies to views with which he agreed, as in fact Robert of Curzon does in his *Summa* (see Example 6.1). If the traditional date of PP's text were correct, it would mean that by 1176 the recursive argument method had been not only translated and received by Western European scholars but already assimilated and regularized. Yet leaving aside the translations of Avicenna, there is absolutely no evidence for anyone else having used the method in Latin texts before ca. 1202 (at the earliest), when it appears in Robert of Curzon's *Summa*. Other contemporaries of PP, not to speak of those of earlier generations, did not know the method at all. This would be absolutely baffling if it were not clear, as Valente and Baldwin have suggested, that the received text has been much reworked by PP himself or, much more probably, by one or more of his students.

It must be concluded that the recursive argument method in PP's *Sentences* does not date to 1176, but rather was added to the text during one or more revisions or rewritings during the early thirteenth century. Therefore, although this conclusion is only provisional pending more work on the topic by specialists, the *Sententiae libri quinque* is evidently not the earliest datable example of the use of the recursive method in native (not translated) Medieval Latin works. That honor continues to belong to Robert of Curzon's *Summa*, an example from which is given and discussed in chapter 6.

It is worth repeating that there remain very many problems connected with the dating, attribution, and interpretation of the works of "Peter of Poitiers." The provisional nature of this appendix is due primarily to the highly problematic modern scholarly views on his life (including the issue of which Peter of Poitiers did what, when, and the question of how many PPs there really were) and the so far inadequate studies of the manuscripts and versions of the *Sentences.* Because the work has hitherto been considered to be a unitary text by a unitary author, firmly dated "as is" (whatever that could mean without a serious, full critical edition), it has been used as the linchpin for the dating, in turn, of most of the other important late twelfth-century theological works and their authors. The authorship and dating of PP and his *Sentences* must therefore be restudied in detail by specialists before any conclusions can be drawn.[30]

## The Recursive Method in Peter of Poitiers' *Sentences*

The example given below is taken from chapter 14 of Book II of the *Sentences.* The editors give a brief outline of the argument:

> After a statement of opinions and the order of treatment (138–27, p. 87) Peter defines will, end, and intention (28–37, p. 88; cf. 16.142–151, p. 117). He then asks whether will is sin and, if it is, whether it is the same sin numerically as the exterior act (XIV). All agree, he says, that the will is sin (14.1–2, p. 89). Why is the will sin rather than other natural powers (7–23, p. 89)? He then takes up the question of the identity of interior and exterior acts, presenting objections and offering answers to both the affirmative and negative opinions (24–120, pp. 90–93). Returning to his first statement, that the will is sin, he brings up two practical objections to the common opinion (121–153, 154-216, pp. 93–96), two objections based upon an *auctoritas* of St. Augustine: "every sin is so far voluntary that if it is not voluntary it is not sin" (217–284, pp. 96–99), and another objection based upon the statement: "all merit depends upon the will" (285–340, pp. 99–101). Definitive solutions he humbly leaves to better minds (341–343, p. 101).[31]

A part of this text that may be analyzed as a recursive argument method is outlined below, but as the editors' content-based outline shows, it is hardly a "classic" example of a recursive argument. Although it might be said that most of the argument in the chapter is actually outside the structure of the recursive argument as given in Example B.1, in

[30] This would seem to be an ideal topic for a doctoral dissertation in medieval studies.

[31] Latin: Moore et al. (1950: 89–101).

fact the boundaries between the recursive argument and other parts of the author's argument are not clear. They cross-refer and intersect each other at several points in ways not seen in the well-known method used in works written slightly later in the thirteenth century. Although other recursive arguments can be similarly isolated in the *Sentences*, they are rarely even as clear-cut as the one examined here; most of the arguments in the first two books cannot be analyzed as recursive arguments at all. In addition, what has been marked in the analysis below as the AUTHOR'S VIEW ARGUMENT (following the explicit separation of that section in the published edition) is not really an argument.

These points further support the argument presented above that the use of the recursive argument method in the *Sentences* is an artifact of a late rewriting of the work, possibly by Peter of Poitiers himself. But because the usage is so irregular—evidently the author or revisor was still not quite clear about how the recursive argument method works—it would also support the view that it is one of the earliest original Latin works to use the method, and might therefore be important evidence for the development of what became the "classic" method used in works written from the mid-thirteenth century on. Nevertheless, in order to attempt to address that issue it is first necessary to undertake a thorough reexamination of the many questions concerning Peter of Poitiers and the *Sententiae libri quinque* that are raised in this appendix.

### *Example B.1*

**PETER OF POITIERS**

***Sententiarum libri quinque***

BOOK II, CHAPTER 14[32]

Whether Willing [evil] is a Sin and, if it is a Sin, Whether it is the Same Sin as an [evil] Action, or a Different [sin]

[I. MAIN ARGUMENT]
. . . Since it is agreed by all that willing [evil] is a sin, it is asked whether willing [evil] is the same sin as the exterior action or not. . . .

[II. SUBARGUMENTS$_1$]
[PRO] To those, therefore, who say that willing [evil] is *the same* sin as the exterior action, . . .

[32] Latin: Moore et al. (1950: viii, 89ff.).

[Pro]

[1] Supposing that someone [both] wishes to kill a person and does kill him; he is considered to be guilty of a mortal sin, "but no sin is admitted . . . "[33]

[2] Or in the same way: there is no more sin in his having killed someone than there was in him before the act; therefore no sin is incurred for his having killed someone.

[3] Again, wishing to kill is not killing, therefore willing homicide is not homicide.

[Contra] Or contrariwise: the exterior act is the same sin as the interior act; the exterior act is homicide; therefore the interior act is homicide . . .

[4] The apostle [Paul] somewhere speaks of *homicides*, in the plural . . .

[5] Again, a different legal commandment forbids the will to steal, namely, "Thou shalt not covet thy neighbor's goods."

[6] Again, it is read in the *Gloss on Exodus* that "whoever adds an evil deed to an evil will adds iniquity to iniquity." But . . .

[Contra] If it is said that an evil will and [evil] act are *different* sins,

[Pro]

[7] then a different punishment must be meted out for the willing and another for the act.

[8] Again, it is because of an evil act that someone is forbidden to take holy orders, never because of willing [evil], therefore the act is a greater sin than the willing.

[Contra]

[9] Contrariwise, willing [evil] is never without contempt, while the act is without contempt; therefore, the willing is a greater sin than the act.

[10] Again, all merit is on account of willing and not on account of acting. . . .

[11] Again, [evil] will is never without sin; but an [evil] act is sometimes without sin . . .

[33] The relevant passage of Peter Lombard's *Sentences* (Book II, Distinction XLII, 1.3; Silano 2008: 207) is quoted here.

[III. AUTHOR'S VIEW ARGUMENT[34]]

It does not matter much whether it is said that the willing and the act are the same sin or different [sins]. We respond therefore both to those which are objected against the first part as well as those against the other [part].[35]

[IV. **SUBARGUMENTS$_2$**]

[PRO]

[PRO]

[1] So, to that which is said first, "He is considered to be guilty of a mortal sin, but no [sin is admitted . . . ]," and so on, we say . . .

[2] [No reply.]

[3] To that which is objected afterward, "Wishing to kill is not killing," and so on, it is said that . . .

[CONTRA]

[4] So too, the apostle says in the plural *homicides* . . .

[5] To that which is said afterward, we say that a different commandment . . .

[6] To what is said afterward: "Iniquity to iniquity," and so on, this is said not because the deed is a different sin from the willing but . . .

[CONTRA]

[PRO]

[7] To what is said after [that], against those who say that the willing [of evil] and the act are different sins . . .

[8] What, however, is said [next]—"it is because of an [evil] act that someone is forbidden to take holy orders, and not because of willing [evil]," and so on—is not valid. . . .

[CONTRA]

[9] [No reply.]

[10] [No reply.]

[11] [No reply.]

[34] This obviously does not read like an argument at all, but rather as an introduction to the SUBARGUMENTS$_2$ section, as remarked above.

[35] Despite this remark (on which see the comments above in this appendix), the text does omit replies to ARGUMENTS [9], [10], and [11].

## Appendix C
# Charter of the Collège des Dix-huit

**50.** ***Fundatio Collegii scholarium decem et octo seu « des Dix-Huit ».***

*1180. Parisiis.*[1]

Ego Barbedaurus, Parisiensis ecclesie decanus, et universum ejusdem ecclesie capitulum. Notum fieri volumus omnibus, tam presentibus quam futuris, quod cum dominus Jocius de Londiniis reversus fuisset Iherosolimis, inspecto summo devotionis affectu beneficio, quod in hospicio Beate Marie Parisiensis pauperibus et egris administrantur, ibidem cameram quandam, in qua pauperes clerici ex antiqua consuetudine hospitabantur, inspexit, et illam a procuratoribus ejusdem domus ad usum predictorum clericorum precio quinquaginta duarum librarum de consilio nostro et magistri Hilduini, Parisiensis cancellarii, ejusdem loci tunc procuratoris, inperpetuum adquisivit, tali facta conditione, quod ejusdem domus procuratores decem et octo scolaribus clericis lectos sufficientes et singulis mensibus duodecim nummos de confraria que colligitur in archa, perpetuo administrabunt. Predictos vero clericos ante corpora in eadem domo defuncta crucem et aquam benedictam secundum vices suas deferre, et singulis noctibus septem psalmos penitentiales et orationes debitas et ex antiquo institutas celebrare oportebit. Ut autem hoc firmum ac stabile maneret, prefatus Jocius hanc cartam nostre constitutionis prefatis clericis fieri impetravit et sigilli nostri caractere subternotato corroborari postulavit. Actum publice Parisius in capitulo nostro, anno ab Incarnatione Domine M° C° octogesimo. S[ignum] Barbedauri decani. S. Galteri precentoris. S. Philippi archidiaconi. S. Graciani archidiaconi. S. Giraldi archidiaconi. S. Galteri presbyteri. S. Jocelini presbyteri. S. Galonis presbyteri. S. Sy[monis] diaconi. S. Petri diaconi. S. Odonis diaconi. S. Bauduini subdiaconi. S. Ade subdiaconi. S. Hugonis subdiaconi. S. Stephani pueri. S. Mauricii pueri. S. Gaufridi pueri. Datum per manum Hilduini cancellarii.

[1] Denifle (1899: 49–50), from an early manuscript copy; the italicized information has been added by the editors.

# REFERENCES

N.B.: I have provided the original language titles only for works that I have cited from the original or from which I have translated passages.


Abelard, Peter. 1976–1977. *Sic et non*. Ed. Blanche B. Boyer and Richard McKeon, *Abailard, Sic et non: A Critical Edition*. Chicago: University of Chicago Press.

Achena, Mohammad, and Henri Massé. 1955. *Le livre de Science,* vol. 1. Paris: Société d'édition 'Les Belles Lettres'.

Adamson, Peter. 2007. *Al-Kindī*. New York: Oxford University Press.

Albertus Magnus. 1955. *Quaestiones super De animalibus*. Ed. Ephrem Filhaut. In Bernhardo Geyer, ed., *Sancti doctoris ecclesiae Alberti Magni Ordinis Fratrum praedicatorum episcopi Opera Omnia*, vol. 12. Aschendorff: Monasterii Westfalorum.

Al-Bīrūnī, Abū al-Raiḥān Muḥammad ibn Aḥmad. 1910. *An Accurate Description of All Categories of Hindu Thought, as Well Those Which Are Admissible as Those Which Must Be Rejected*. Trans. Edward C. Sachau. *Alberuni's India: An Account of the Religion, Philosophy, Literature, Geography, Chronology, Astronomy, Customs, Laws and Astrology of India about ad 1030*. London: Kegan Paul, Trench, Trubner. Repr. New Delhi: Oriental Books Reprint Corporation, 1983.

Alexander of Hales. 1924. *Summa theologica*. Ed. Collegio di S. Bonaventura. *Doctoris irrefragabilis Alexandri de Hales Ordinis Minorum Summa theologica*, vol. 1. Quaracchi: Ex Typographia Collegii S. Bonaventurae.

———. 1960. *Quaestiones disputatae "Antequem esset frater,"* vol. 2. Quaracchi: Ex Typographia Collegii S. Bonaventurae.

Al-Ghazālī, Abū Ḥamīd Muḥammad. 1953. *Deliverance from Error and Attachment to the Lord of Might and Majesty*. Trans. M. Montgomery Watt. In *The Faith and Practice of al-Ghazálí*, 17–92. London: George Allen and Unwin.

———. 2000. تهافت الفلاسفة *The Incoherence of the Philosophers*. Ed. and trans. Michael E. Marmura. 2nd edition. Provo, UT: Brigham Young University Press.

Al-Jūzjānī, Abū 'Ubayd. 1974. سيرة الشيخ الرئيس Ed. and trans. William E. Gohlman. *The Life of Ibn Sina*. Albany: State University of New York Press.

Almor, Amit. 2008. Why Does Language Interfere with Vision-Based Tasks? *Experimental Psychology* 55, no. 4: 260–268.

Al-Mūsawī: *See* Al-Rāzī, Fakhr al-Dīn.

Al-Rāzī, Fakhr al-Dīn. 1934. كتاب الاربعين في اصول الدين Ed. Sayyid Zayn al-'Ābidīn al-Mūsawī. Hyderabad: Majlis Dā'ira al-Ma'ārif al-'Uthmāniyya.

d'Alverny, Marie-Thérèse 1952. Notes sur les traductions médiévales d'Avicenne. *Archives d'Histoire Doctrinale et Littéraire du Moyen Age* 27: 337–366.

———. 1954. Avendauth? In *Homenaje a Millás-Vallicrosa*, vol. 1: 19–43. Barcelona: Consejo Superior de Investigaciones Científicas.

———. 1961. Avicenna latinus. *Archives d'histoire doctrinale et littéraire du moyen âge* 36: 281–324.

———. 1982. Translations and Translators. In Robert L. Benson, Giles Constable, and Carol D. Lanham, eds., *Renaissance and Renewal in the Twelfth Century*, 421–462. Cambridge, MA: Harvard University Press. Repr. Toronto: University of Toronto Press, 1991.

———. 1989. Les traductions a deux interprètes: d'arabe en langue vernaculaire et de langue vernaculaire en latin. In *Traduction et traducteurs au moyen âge. Actes du colloque international du CNRS organisé à Paris, Institut de recherche et d'histoire des textes, les 26–28 mai 1986*, 193–206. Paris: Éditions du Centre National de la Recherche Scientifique.

———. 1991. Translations and Translators. In Robert L. Benson, Giles Constable, and Carol D. Lanham, eds., *Renaissance and Renewal in the Twelfth Century*, 421–462. Toronto: University of Toronto Press.

American Psychological Association. 1983. *Publication Manual of the American Psychological Association*. 3rd edition. Washington, DC: American Psychological Association.

Anawati: *See* Avicenna.

Aquinas, Thomas. 1975. *Summa Theologiæ*. Ed. and trans. Marcus Lefébure. In *St. Thomas Aquinas, Summa Theologiæ*, vol. 38, *Injustice*. London: Blackfriars and Eyre & Spottiswoode.

Aristotle. 1984. *The Complete Works of Aristotle: The Revised Oxford Translation*. Ed. Jonathan Barnes. Princeton, NJ: Princeton University Press.

Arqués, Rossend, and Raffaele Pinto. 1995. La invenció del sonet i la cultura a la Sicília de Frederic II. *L'Avenç, revista d'Història* 195: 34–37.

Association Guillaume Budé. 1972. *Règles et recommandations pour les éditions critiques (Série grecque)*. Paris: Les belles lettres.

Avicenna (Ibn Sīnā, Abū ʿAlī al-Ḥusain). 1956. الفن السادس من الطبيعيات (علم النفس) من كتاب الشفاء (*Psychologie*). Trans. Ján Bakoš. *Psychologie d'Ibn Sina (Avicenne) d'après son oeuvre aš-Šifāʾ*. Prague: Académie Tchécoslavaque des Sciences.

———. 1960. الشفاء : الإلهيات (*La Métaphysique*). Vol. 1, ed. G. C. Anawati and Saʿid Zayed; vol. 2, ed. Mohammed Youssef Moussa, Solayman Dunya, and Saʿid Zayed. Cairo: Organisation Générale des Imprimeries Gouvernementales.

———. 1968. *Avicenna Latinus: Liber de anima seu sextus de naturalibus, iv-v*. Ed. Simone Van Riet. Louvain: E. Peeters.

———. 1972. *Avicenna Latinus: Liber de anima seu sextus de naturalibus, i-ii-iii*. Ed. Simone Van Riet. Louvain: E. Peeters.

———. 1977. *Avicenna Latinus: Liber de philosophia prima sive scientia divina, i–iv*. Ed. Simone Van Riet. Louvain: E. Peeters.

———. 1980. *Avicenna Latinus: Liber de philosophia prima sive scientia divina, v–x*. Ed. Simone Van Riet. Louvain: E. Peeters.

———. 1985. *La Métaphysique du Shifāʾ*, vol. 2 (books vi to x). Trans. Georges C. Anawati. Paris: Librairie Philosophique J. Vrin.

———. 2005. الشفاء : الإلهيات *The Metaphysics of* The Healing. Ed. and trans. Michael E. Marmura. Provo, UT: Brigham Young University Press.

Bacon, Roger. 1930. *Questiones supra libros prime philosophie Aristotelis* (METAPHYSICA i, ii, v–x). Ed. Robert Steele. Opera hactenus inedita Rogeri Baconi Fasc. x. Oxford: Clarendon.

———. 1935. *Questiones supra libros octo* Physicorum *Aristotelis*. Ed. Ferdinand M. Delorme. Opera hactenus inedita Rogeri Baconi Fasc. xiii. Oxford: Clarendon.

Bakoš: *See* Avicenna.

Baldwin, John W. 1970. *Masters, Princes, and Merchants: The Social Views of Peter the Chanter and His Circle*, vol. 1. Princeton, NJ: Princeton University Press.

Barnes: *See* Aristotle.

Barthold, W. 1964. Сочинения, II, 2. Moscow: Izdatel'stvo vostochnoj literatury.

Bazàn, Bernardo C. 1985. Les questions disputées, principalement dans les facultés de theologie. In Bernardo Bazàn, John W. Wippel, Gérard Fransen, and Danielle Jacquart, *Les questions disputées et les questions quodlibétiques dans les facultés de théologie, de droit et de médicine*, 15–149. Typologie des sources du moyen âge occidental, fasc. 44–45. Brepols: Turnhout.

Bazàn, Bernardo, John W. Wippel, Gérard Fransen, and Danielle Jacquart. 1985. *Les questions disputées et les questions quodlibétiques dans les facultés de théologie, de droit et de médicine*. Typologie des sources du moyen âge occidental, fasc. 44–45. Brepols: Turnhout.

Beckwith, Christopher I. 1979. The Introduction of Greek Medicine into Tibet in the Seventh and Eighth Centuries. *Journal of the American Oriental Society* 99: 297–3l3.

———. 1984a. Aspects of the Early History of the Central Asian Guard Corps in Islam. *Archivum Eurasiae Medii Aevi* 4: 29–43. Reprinted in C. Edmund Bosworth, ed., *The Turks in the Early Islamic World* (Aldershot: Ashgate, 2007), 275–289.

———. 1984b. The Plan of the City of Peace: Central Asian Iranian Factors in Early 'Abbâsid Design. *Acta Orientalia Academiae Scientiarum Hungaricae* 38: 128–147.

———. 1990. The Medieval Scholastic Method in Tibet and the West. In L. Epstein and R. Sherburne, eds., *Reflections on Tibetan Culture: Essays in Memory of Turrell V. Wylie*, 307–313. Lewiston, ME: E. Mellen Press.

———. 1993. *The Tibetan Empire in Central Asia: A History of the Struggle for Great Power among Tibetans, Turks, Arabs, and Chinese during the Early Middle Ages*. Rev. edition. Princeton, NJ: Princeton University Press.

———. 2007a. *Koguryo, the Language of Japan's Continental Relatives: An Introduction to the Historical-Comparative Study of the Japanese-Koguryoic Languages, with a Preliminary Description of Archaic Northeastern Middle Chinese*. 2nd edition. Leiden: Brill.

———. 2007b. *Phoronyms: Classifiers, Class Nouns, and the Pseudopartitive Construction*. New York: Peter Lang.

———. 2009. *Empires of the Silk Road: A History of Central Eurasia from the Bronze Age to the Present*. Princeton, NJ: Princeton University Press.

———. 2010a. Could There Be a Korean–Japanese Linguistic Relationship Theory? Science, the Data, and the Alternatives. *International Journal of Asian Studies* 7, no. 2: 201–219.

———. 2010b. The Sarvāstivādin Buddhist Scholastic Method in Medieval Islam and Tibet. In Anna Akasoy, Charles Burnett, and Ronit Yoeli-Tlalim, eds., *Islam and Tibet: Interactions along the Musk Routes*, 163–175. Farnham: Ashgate.

———. 2011. Pyrrho's Logic. *Elenchos* 32, no. 2: 287–327.

Bernard, Paul. 1978. Campagne de Fouilles 1976–1977 à Ai Khanoum (Afghanistan). *Comptes Rendu de l'Académie des Inscriptions et Belles Lettres* 122, no. 2: 421–463.

Bhattacharya, Kamaleswar, E. H. Johnston, and Arnold Kunst. 1978. *The Dialectical Method of Nāgārjuna*. Delhi: Motilal Banarsidass.

Bieniak, Magdalena. 2010. *The Soul-Body Problem at Paris, ca. 1200–1250: Hugh of St.-Cher and His Contemporaries*. Leuven: Leuven University Press.

Bosworth, Clifford Edmund. 2008. The Appearance and Establishment of Islam in Afghanistan. In Étienne de la Vaissière, ed., *Islamisation de l'Asie centrale: Processus locaux d'acculturation du vii^e au xi^e siècle*, 97–114. Paris: Association pour l'Avancement des Études Iraniennes.

Boyer and McKeon: *See* Abelard.

Brooks, E. Bruce. 1999. *Alexandrian Motifs in Chinese Texts*. Sino-Platonic Papers 96. Department of East Asian Languages and Civilizations, University of Pennsylvania.

Bulliet, Richard W. 1972. *The Patricians of Nishapur: A Study in Medieval Islamic Social History*. Cambridge, MA: Harvard University Press.

———. 1976. Naw Bahār and the Survival of Iranian Buddhism. *Iran: Journal of the British Institute of Persian Studies* 14: 140–145.

Burnett, Charles. 1998. Islamic Philosophy: Transmission into Western Europe. In E. Craig, ed., *Routledge Encyclopedia of Philosophy*. London: Routledge. Retrieved January 9, 2011, from www.rep.routledge.com/article.

Ch'en, Mei-Chin. 1992. The Eminent Chinese Monk Hsuan-Tsang: His Contributions to Buddhist Scripture Translation and to the Propagation of Buddhism in China. Ph.D. dissertation, University of Wisconsin, Madison.

Cox, Collett. 1995. *Disputed Dharmas: Early Buddhist Theories on Existence. An Annotated Translation of the Section on Factors Dissociated from Thought from Saṅghabhadra's Nyāyānusāra*. Tokyo: International Institute for Buddhist Studies.

———. 1998. Chapter Three: Kaśmīra: Vaibhāṣika Orthodoxy. In Charles Willemen, Bart Dessein, and Collett Cox, *Sarvāstivāda Buddhist Scholasticism*, 138–254. Leiden: Brill.

Crombie, A. C. 1953. *Robert Grosseteste and the Origins of Experimental Science, 1100–1700*. Oxford: Clarendon Press.

De Jong, J. W. 1993. The Beginnings of Buddhism. *The Eastern Buddhist* N.S. 26, no. 2: 11–30.

de la Vaissière, Étienne. 2007. *Samarcande et Samarra: Élites d'Asie centrale dans l'empire abbasside*. Paris: Association pour l'avancement des études iraniennes.

———. 2008. Le *Ribāṭ* d'Asie centrale. In Étienne de la Vaissière, ed., *Islamisation de l'Asie centrale: Processus locaux d'acculturation du vii^e au xi^e siècle*, 71–94. Paris: Association pour l'Avancement des Études Iraniennes.

Denifle, Henricus. 1899. *Chartularium Universitatis Parisiensis*. Paris; repr. Brussels: Culture et Civilisation, 1964.

Derge: *See* Kamalaśīla.

Dessein: *See* Dharmatrāta.

Dhanani, Alnoor. 1994. *The Physical Theory of Kalām: Atoms, Space, and Void in Basrian Muʿtazilī Cosmology*. Leiden: Brill.

Dharmatrāta. 1999. *Saṃyuktābhidharmahṛdaya: Heart of Scholasticism with Miscellaneous Additions*. Ed. and trans. Bart Dessein. 3 parts. Delhi: Motilal Banarsidass.

Diogenes Laertius. 1925. *Lives of Eminent Philosophers*. Ed. and trans. R. D. Hicks. 2 vols. Cambridge, MA: Harvard University Press. Repr. 1979.

Dreyfus, Georges B. 2008. What Is Debate For? The Rationality of Tibetan Debates and the Role of Humor. *Argumentation* 22: 43–58.

Druart, Thérèse-Anne. 2003. Philosophy in Islam. In A. S. McGrade, ed. *The Cambridge Companion to Medieval Philosophy*, 97–120. Cambridge: Cambridge University Press.

Duri, A. A. 1960. Baghdād. In *E.I.*$_2$ 1: 894–908.

Dutt, Sukumar. 1962. *Buddhist Monks and Monasteries of India: Their History and Their Contribution to Indian Culture.* London: George Allen & Unwin.

Duyvendak, J.J.L. 2001. Paul Pelliot. In Hartmut Walravens, *Paul Pelliot (1878–1945): His Life and Works—A Bibliography*, xiii–xxiv. Bloomington: Indiana University, Research Institute for Inner Asian Studies.

Eastwood, Bruce S. 2002. *The Revival of Planetary Astronomy in Carolingian and Post-Carolingian Europe*. Aldershot: Ashgate Variorum.

El-Rouayheb, Khaled. 2010. *Relational Syllogisms and the History of Arabic Logic, 900–1900*. Leiden: Brill.

Fakhry, Majid. 1958. *Islamic Occasionalism, and Its Critique by Averroës and Aquinas*. London: George Allen & Unwin.

———. 1983. *A History of Islamic Philosophy*. 2nd edition. New York: Columbia University Press.

Ferruolo, Stephen C. 1985. *The Origins of the University: The Schools of Paris and Their Critics, 1100–1215*. Stanford, CA: Stanford University Press.

Fidora, Alexander. 2003. *Die Wissenschaftstheorie des Dominicus Gundissalinus: Voraussetzungen und Konsequenzen des zweiten Anfangs der aristotelischen Philosophie im 12. Jahrhundert*. Berlin: Akademie Verlag.

———. 2004. Dominicus Gundissalinus und die arabische Wissenschaftstheorie. In Andreas Speer and Lydia Wegener, eds., *Wissen über Grenzen: Arabisches Wissen und lateinisches Mittelalter*, 467–482. Berlin: Walter de Gruyter.

Filhaut: *See* Albertus Magnus.

Floridi, Luciano. 2002. *Sextus Empiricus: The Transmission and Recovery of Pyrrhonism.* Oxford: Oxford University Press.

———. 2010. The Rediscovery and Posthumous Influence of Scepticism. In Richard Bett, ed., *The Cambridge Companion to Ancient Scepticism*, 267–287. Cambridge: Cambridge University Press.

Franco, Eli. 2004. *The Spitzer Manuscript: The Oldest Philosophical Manuscript in Sanskrit*. Vienna: Verlag der Österreichischen Akademie der Wissenschaften.

Fransen, Gérard. 1985. Les questions disputées dans les facultés de droit. In Bernardo Bazàn, John W. Wippel, Gérard Fransen, and Danielle Jacquart, *Les questions disputées et les questions quodlibétiques dans les facultés de théologie, de droit et de médicine*, 223–277. Typologie des sources du moyen âge occidental, fasc. 44–45. Brepols: Turnhout.

Frauwallner, Erich. 1957. The Historical Data We Possess on the Person and the Doctrine of the Buddha. *East and West* 7: 309–312. Reprinted in E. Frauwallner, *Kleine Schriften*, ed. Ernst Steinkellner (Wiesbaden: Steiner, 1982), 703ff.

Fuchs, Walter. 1938. Hui-ch'ao's Pilgerreise durch Nordwest-Indien und Zentral-Asien um 726. *Sitzungsberichte der Preussischen Akademie der Wissenschaften (philosophisch-historische Klasse)* 30: 426–469.

Galilei, Galileo. 1967. *Dialogue Concerning the Two Chief World Systems.* Trans. Stimson Drake. Berkeley: University of California Press.

Gethin, Rupert. 1998. *The Foundations of Buddhism*. Oxford: Oxford University Press.

Gimaret, D. 1993. Muʿtazila. In *E.I.*$_2$ 7: 783–793.

Glorieux, P. 1952. Le *Contra quatuor labyrinthos Franciae* de Gauthier de Saint-Victor: Édition critique. *Archives d'Histoire Doctrinale et Littéraire du Moyen Age* 19: 188–335.

Gohlman, *See* al-Jūzjānī.

Gohlman, William E. 1974. *The Life of Ibn Sina: A Critical Edition and Annotated Translation*. Albany: State University of New York Press.

Goodman, Lenn E. 2006. *Avicenna*. Updated edition. Ithaca, NY: Cornell University Press.

Goodwin, C. James. 1998. *Research in Psychology: Methods and Design*. 2nd edition. New York: John Wiley and Sons.

Gower, Barry. 1997. *Scientific Method: An Historical and Philosophical Introduction*. London: Routledge.

Grabmann, Martin. 1909–1911. *Die Geschichte der scholastischen Methode*. 2 vols. Freiburg im Bresgau: Herder. Repr. Graz: Akademische Druck- und Verlags-anstalt, 1957.

Grant, Edward. 1974. *A Source Book in Medieval Science*. Cambridge, MA: Harvard University Press.

———. 1996. *The Foundations of Modern Science in the Middle Ages*. Cambridge: Cambridge University Press.

———. 2004. *Science and Religion, 400 B.C. to A.D. 1550: From Aristotle to Copernicus*. Westport, CT: Greenwood Press.

———. 2007. *A History of Natural Philosophy: From the Ancient World to the Nineteenth Century*. Cambridge: Cambridge University Press.

———. 2009. Let's Pretend: Imagination and the Birth of Science Fiction in Medieval Europe. Unpublished lecture, Sept. 18, 2009, Indiana University, Bloomington.

Gutas, Dimitri. 1988. *Avicenna and the Aristotelian Tradition: Introduction to Reading Avicenna's Philosophical Works*. Leiden: Brill.

———. 1998. *Greek Thought, Arabic Culture: The Graeco-Arabic Translation Movement in Baghdad and Early ʿAbbāsid Society (2nd–4th/8th–10th Centuries)*. London: Routledge.

———. 2010. Origins in Baghdad. In Robert Pasnau, ed., *The Cambridge History of Medieval Philosophy*, vol. 1: 11–25. Cambridge: Cambridge University Press.

Halkias, Georgios. 2010. The Muslim Queens of the Himalayas: Princess Exchanges in Baltistan and Ladakh. In Anna Akasoy, Charles Burnett, and Ronit Yoeli-Tlalim, eds., *Islam and Tibet: Interactions along the Musk Routes*, 231–252. Farnham: Ashgate.

Häring, Nicholas M. 1962. The Porretans and the Greek Fathers. *Mediaeval Studies* 24: 181–209.

Hartmann, Jens-Uwe. 1999. Buddhist Sanskrit Texts from Northern Turkestan and Their Relation to the Chinese Tripiṭaka. In *Buddhism across Boundaries: Chinese Buddhism and the Western Regions* [= *Collection of Essays 1993*], 108–136. Taipei: Fo Guang Shan Foundation for Buddhist and Culture Education.

Haskins, Charles Homer. 1927. *The Renaissance of the Twelfth Century*. Cambridge, MA: Harvard University Press.

Hasse, Dag Nikolaus. 2004. The Social Conditions of the Arabic-(Hebrew)-Latin Translation Movements in Medieval Spain and in the Renaissance. In Andreas Speer and Lydia Wegener, eds., *Wissen über Grenzen: Arabisches Wissen und lateinisches Mittelalter*, 68–86. Berlin: Walter de Gruyter.

Hauréau, B. 1890. *Notices et extraits de quelques manuscrits Latins de la Bibliothèque Nationale*. Vol. 1. Paris: Librairie C. Klincksieck.

Herrmann-Pfandt, Adalheid. 2008. *Die Lhan kar ma: Ein früher Katalog der ins Tibetische übersetzten buddhistischen Texte*. Vienna: Verlag der Österreichischen Akademie der Wissenschaften (= Österreichische Akademie der Wissenschaften, Philosophisch-historische Klasse, Denkschriften, 367, Band).

Hillenbrand, R. 1986. III. Architecture. In Pedersen et al., Madrasa. *E.I.*$_2$ 5: 1136–1154.

Hirakawa, Akira. 1990. *A History of Indian Buddhism: from Śākyamuni to Early Mahāyāna*. Trans. Paul Groner. Honolulu: University of Hawaii Press.

Hsüan Tsang. 2000. 大唐西域記校注. Ed. Chi Hsien-Lin et al. Peking: Chung-hua shu-chü.

Ibn Sīnā, Abū 'Alī al-Ḥusain: *See* Avicenna.

Jackson, David P. 1987. *The Entrance Gate for the Wise (Section III): Sa-skya Paṇḍita on Indian and Tibetan Traditions of Pramāṇa and Philosophical Debate*. Vienna: Arbeitskreis für Tibetische und Buddhistische Studien, Universität Wien.

Jacobi, Klaus. 1998. Gilbert of Poitiers. In E. Craig, ed., *Routledge Encyclopedia of Philosophy*. London: Routledge. Retrieved January 9, 2011, from www.rep.routledge.com/article.

Jolivet, Jean. 1988. The Arabic Inheritance. In Peter Dronke, ed., *A History of Twelfth-Century Western Philosophy*, 113–148. Cambridge: Cambridge University Press.

Kamalaśīla. 1985. དབུ་མ་སྣང་བ་ [= *Madhyamakāloka*]. Sde-dge Bstan-'gyur series, vol. 107. Dbu ma'i skor, vol. 12. Delhi: Delhi Karmapae Choedhey.

Kantorowicz, Ernst H. 1938. The *Quaestiones Disputatae* of the Glossators. *Revue d'histoire du droit* 16: 1–67.

Keira, Ryusei. 2004. *Mādhyamika and Epistemology: A Study of Kamalaśīla's Method for Proving the Voidness of All Dharmas*. Vienna: Arbeitskreis für Tibetische und Buddhistische Studien, Universität Wien.

King, Anya. 2007. The Musk Trade and the Near East in the Early Medieval Period. Ph.D. dissertation, Indiana University, Bloomington.

———. 2010. Tibetan Musk and Medieval Arab Perfumery. In Anna Akasoy, Charles Burnett, and Ronit Yoeli-Tlalim, eds., *Islam and Tibet: Interactions along the Musk Routes*, 145–161. Farnham: Ashgate.

King, David A. 2008. Mathematical Geography in Fifteenth-Century Egypt: An Episode in the Decline of Islamic Science. In Anna Akasoy and Wim Raven, eds., *Islamic Thought in the Middle Ages: Studies in Text, Transmission and Translation, in Honour of Hans Daiber*, 319–344. Leiden: Brill.

Klimburg-Salter, Deborah. 2008. Buddhist Painting in the Hindu Kush ca. VIIth to Xth Centuries: Reflections on the Co-existence of Pre-Islamic and Islamic Artistic Cultures during the Early Centuries of the Islamic Era. In Étienne de la Vaissière, ed., *Islamisation de l'Asie centrale: Processus locaux d'acculturation du vii*$^{e}$ *au xi*$^{e}$ *siècle*, 131–159. Paris: Association pour l'Avancement des Études Iraniennes.

Kritzeck, James. 1964. *Peter the Venerable and Islam*. Princeton, NJ: Princeton University Press.

Kritzer, Robert. 2005. *Vasubandhu and the Yogācārabhūmi: Yogācāra Elements in the Abihidharmakośabhāṣya*. Tokyo: International Institute for Buddhist Studies.

Lamotte, Étienne. 1988. *History of Indian Buddhism*. Louvain-la-neuve: Université Catholique de Louvain.

Lassner, Jacob. 1970. *The Topography of Baghdad in the Early Middle Ages: Text and Studies*. Detroit: Wayne State University Press.

Lawn, Brian. 1993. *The Rise and Decline of the Scholastic 'quaestio disputata'*. Leiden: Brill.

Lefébure: *See* Aquinas.

Lefèvre, Georges. 1902. *Le traité* "De usura" *de Robert de Courçon*. Lille: Travaux et mémoires de l'Université de Lille, vol. 10, mémoire № 30.

Lerner, Jeffrey D. 2003. The Aï Khanoum Philosophical Papyrus. *Zeitschrift für Papyrologie und Epigraphik* 142: 45–51.

Li, Rongxi, trans. 1995. *A Biography of the Tripiṭaka Master of the Great Ci'en Monastery of the Great Tang Dynasty.* Berkeley, CA: Numata Center for Buddhist Translation and Research.

Lindberg, David C. 1976. *Theories of Vision from al-Kindī to Kepler*. Chicago: University of Chicago Press.

———. 2007. *The Beginnings of Western Science: The European Scientific Tradition in Philosophical, Religious, and Institutional Context, Prehistory to a.d. 1450*. 2nd edition. Chicago: University of Chicago Press.

Little, A. G., and F. Pelster. 1934. *Oxford Theology and Theologians c. A.D. 1282–1302*. Oxford: Clarendon Press.

Litvinskij, Boris A. 1985. Ajina Tepe. In Ehsan Yarshater, ed., *Encylopaedia Iranica*, vol. 1: 703–705. London: Routledge & Kegan Paul.

Litvinskij, Boris A., and Tamara I. Zeimal'. 1971. Аджина-Тепа. Moscow: Iskusstvo.

Macy, Gary. 2009. A Guide to Thirteenth Century Theologians. University of San Diego. http://home.sandiego.edu/~macy/Robert%20of%20Courson.html, accessed November 2009.

Makdisi, George. 1981. *The Rise of Colleges: Institutions of Learning in Islam and the West*. Edinburgh: Edinburgh University Press.

———. 1990. *The Rise of Humanism in Classical Islam and the Christian West, with Special Reference to Scholasticism*. Edinburgh: Edinburgh University Press.

Marenbon, John. 2002. Gilbert of Poitiers. In Jorge J. E. Gracia and Timothy B. Noone, eds., *A Companion to Philosophy in the Middle Ages*, 264–265. Oxford: Blackwell.

Marmura: *See* Avicenna *and* Al-Ghazālī.

Matar, Nabil I. 2003. *In the Lands of the Christians: Arabic Travel Writing in the Seventeenth Century*. New York: Routledge.

———. 2009. *Europe through Arab Eyes, 1578–1727*. New York: Columbia University Press.

McCluskey, Stephen C. 1998. *Astronomies and Cultures in Early Medieval Europe*. Cambridge: Cambridge University Press.

McEvilley, Thomas. 2002. *The Shape of Ancient Thought: Comparative Studies in Greek and Indian Philosophies*. New York: Allworth Press.

Moore, Philip S. 1936. *The Works of Peter of Poitiers, Master in Theology and Chancellor of Paris (1193–1205)*. Notre Dame, IN: University of Notre Dame.

Moore and Dulong: *See* Peter of Poitiers.

Moore et al.: *See* Peter of Poitiers.

Murdoch, John F. 1974. Logic. In Edward Grant, ed., *A Source Book in Medieval Science*, 77–89. Cambridge, MA: Harvard University Press.

Nallino, C. A. 1986. Al-Battānī. In *E.I.*$_2$ 1: 1104–1105.

Nasr, Seyyed Hossein. 2006. *Islamic Philosophy from Its Origin to the Present: Philosophy in the Land of Prophecy*. Albany: State University of New York Press.

Needham, Joseph. 1954–2008. *Science and Civilization in China*. Cambridge: Cambridge University Press.

Nielsen, Lauge Olaf. 1982. *Theology and Philosophy in the Twelfth Century: A Study of Gilbert Porreta's Thinking and the Theological Expositions of the Doctrine of the Incarnation during the Period 1130–1180*. Leiden: Brill.

Pedersen, J., and G. Makdisi. 1986. I. The Institution in the Arabic, Persian and Turkish Lands. In Pedersen et al., Madrasa, in *E.I.*$_2$ 5: 1123–1134.

Pedersen, J., et al. 1986. Madrasa. In *E.I.*$_2$ 5: 1123–1154.

Pelster: *See* Little and Pelster.

Peter of Poitiers. 1943. *Sententiarum libri quinque*. [I.] Philip S. Moore and Marthe Dulong, eds., *Sententiae Petri Pectaviensis, I*. Notre Dame, IN: University of Notre Dame.

———. 1950. *Sententiarum libri quinque*. [II.] Philip S. Moore, Joseph N. Garvin, and Marthe Dulong, eds., *Sententiae Petri Pectaviensis, II*. Notre Dame, IN: University of Notre Dame.

Pines, Shlomo. 1936. *Beiträge zur Islamischen Atomenlehre*. Berlin: A. Heine.

———. 1997. *Studies in Islamic Atomism*. Trans. Michael Schwarz. Jerusalem: Magnes Press.

Plassen, Jörg. 2002. Die Spuren der Abhandlung (Lun-chi): Exegese und Übung im San-lun des sechsten Jahrhunderts. Ph.D. dissertation, Universität Hamburg.

———. 2007. Das Kŭmgang sammaegyŏng non (T. 1730.34.961a–1008a)—Ein sinokoreanischer buddhistischer Kommentar des 7. Jahrhunderts. In Michael Quisinsky and Peter Walter, eds., *Kommentarkulturen: Die Auslegung zentraler Texte der Weltreligionen, Ein vergleichender Überblick*, 115–134. Cologne: Böhlau Verlag.

Potter, Karl H. 1977. *Encyclopedia of Indian Philosophies: Indian Metaphysics and Epistemology. The Tradition of Nyāya-Vaiśeṣika up to Gaṅgeśa*. Princeton, NJ: Princeton University Press.

Pradhan: *See* Vasubandhu.

Quaracchi: *See* Alexander of Hales.

Pruden: *See* Vasubandhu.

Rapin, Claude. 1992. *Fouilles d'Aï Khanoum VIII: La Trésorerie du Palais Hellénistique d'Aï Khanoum, l'Apogée et la Chute du Royaume Grec de Bactriane*. Paris: de Boccard.

Rashdall, Hastings. 1936. *The Universities of Europe in the Middle Ages*. Ed. F. M. Powicke and A. B. Emden. Vol. 1, *Salerno—Bologna—Paris*. New edition. Oxford: Clarendon. Repr. 1987.

Ritter, H. 2006. Abū Yazīd (Bāyazīd) Ṭayfūr b. ʿĪsā b. Surūshān al-Bisṭāmī. In *E.I.*$_2$ 1: 162, online edition.

Rowson, E. K. 2011. Shubha. In *Encyclopaedia of Islam,* ed. P. Bearman, Th. Bianquis, C. E. Bosworth, E. van Donzel, and W. P. Heinrichs. 2nd edition. Leiden: Brill. Online edition.

Rucquoi, Adeline. 1999. Gundisalvus ou Dominicus Gundisalvi? *Bulletin de philosophie médiévale* 41: 85–106.

Sachau: *See* al-Bīrūnī.

Salomon, Richard, Frank R. Allchin, and Mark Barnard. 1999. *Ancient Buddhist Scrolls from Gandhāra: The British Library Kharoṣṭhī Fragments.* Seattle: University of Washington Press.

Sander, Lore. 1999. Early Prakrit and Sanskrit Manuscripts from Xinjiang (second to fifth/sixth centuries C.E.): Paleography, Literary Evidence, and Their Relation to Buddhist Schools. In *Buddhism across Boundaries: Chinese Buddhism and the Western Regions* [= *Collection of Essays 1993*], 61–106. Taipei: Fo Guang Shan Foundation for Buddhist and Culture Education.

Sayili, Aydin. 1991. Gondēshāpūr. In *E.I.*$_2$ 2: 1119–1120.

Serres, Michel, ed. 1995. *A History of Scientific Thought: Elements of a History of Science.* Oxford: Blackwell.

Sezgin, Fuat. 1970. *Geschichte des arabischen Schrifttums,* vol. 3, *Medizin, Pharmazie, Zoologie, Tierheilkunde bis ca. 430 h.* Leiden: Brill.

———. 1971. *Geschichte des arabischen Schrifttums,* vol. 4, *Alchimie—Chemie—Botanik—Agriculture bis ca. 430 h.* Leiden: Brill.

———. 1974. *Geschichte des arabischen Schrifttums,* vol. 5, *Mathematik bis ca. 430 h.* Leiden: Brill.

———. 1978. *Geschichte des arabischen Schrifttums,* vol. 6, *Astronomie bis ca. 430 h.* Leiden: Brill.

———. 1984. *Geschichte des arabischen Schrifttums,* vol. 9, *Grammatik bis ca. 430 h.* Leiden: Brill.

Shapin, Steven. 1996. *The Scientific Revolution.* Chicago: University of Chicago Press.

Silano, Giulio, trans. 2007. *Peter Lombard, the* Sentences, book 1, *The Mystery of the Trinity.* Toronto: Pontifical Institute of Mediaeval Studies.

———, trans. 2008. *Peter Lombard, the* Sentences, book 2, *On Creation.* Toronto: Pontifical Institute of Mediaeval Studies.

Southern, Richard W. 1995. *Scholastic Humanism and the Unification of Europe,* vol. 1. Oxford: Blackwell.

Starr, S. Frederick. 2009. Rediscovering Central Asia. *Wilson Quarterly* 33, no. 3: 33–43.

Stein, R. A. 1972. *Tibetan Civilization,* trans. J. Driver. Stanford, CA: Stanford University Press.

Summers, William C. 2003. Antibiotics. In J. L. Heilbrun et al., eds., *The Oxford Companion to the History of Modern Science,* vol. 35. New York: Oxford University Press.

Szlezák, Thomas A. 2002. Academy. In *Brill's New Pauly: Encyclopaedia of the Ancient World,* vol. 1: 41–45. Leiden: Brill.

Takakusu, Junjirō, Watanabe Kaigyoku, et al. 1924–1932. 大正新修大藏經. Tokyo: Taishō Issaikyō Kankōkai. (= *Taishō*)

Takeda, Hiromichi, and Collett Cox. 2010. Existence in the Three Time Periods: **Abhidharmamahāvibhāṣāśāstra* (T.1545 pp.393a9–396b23) English Translation.

In Dr. Ronald Y. Nakasone Festschrift Committee, ed., *Memory and Imagination: Essays and Explorations in Buddhist Thought and Culture*, 133–186. Kyoto: Nagata Bunshodo.

Thomas, D. 2000. Al-Ṭabarī, ʿAlī b. Rabban. In *E.I.*$_2$ 10: 17–18.

Thompson, Augustine, James Gordley, and Katherine Christensen. 1979. *The Treatise on Laws*. Washington, DC: Catholic University of America Press.

Thompson, P. M. 1979. *The* Shen Tzu *Fragments*. Oxford: Oxford University Press.

Turner, William. 1907. Anicius Manlius Severinus Boethius. In *The Catholic Encyclopedia*, vol. 2; online edition at www. newadvent.org/cathen/02610b.htm accessed December 12, 2009.

Valente, Luisa. 2007. *Logique et théologie: Les écoles parisiennes entre 1150 et 1200*. Paris: J. Vrin.

Van Bladel, Kevin. 2010. The Bactrian Background of the Barmakids. In Anna Akasoy, Charles Burnett, and Ronit Yoeli-Tlalim, eds., *Islam and Tibet: Interactions along the Musk Routes*, 43–88. Farnham: Ashgate.

Van Ess, Josef. 2002. 60 Years After: Shlomo Pines's Beiträge and Half a Century of Research on Atomism in Islamic Theology. *Proceedings of the Israel Academy of Sciences and Humanities* 8, no. 2: 19–41.

Van Riet: *See* Avicenna.

Vasubandhu. 1975. अभिधर्मकोशभाष्यम्. Ed. P. Pradhan et al., *Abhidharmakośabhāṣyam*. 2nd edition. Patna: K. P. Jayaswal Research Institute.

———. 1989. *Abhidharmakośabhāṣya*. Trans. Leo Pruden, *Abhidharmakośabhāṣyam, by Louis de La Vallée Poussin*. Berkeley, CA: Asian Humanities Press.

Vaux, R. de. 1934. *Notes et textes sur l'avicennisme latin aux confins des XII$^e$–XIII$^e$ siècles*. Paris: J. Vrin.

Verger, Jacques. 1986. *Histoire des universités en France*. Toulouse: Bibliothèque historique Privat.

———. 1995. *Les universités françaises au Moyen Age*. Leiden: Brill.

Vernet, J. 1986. Ibn al-Ḥaytham. In *E.I.*$_2$ 3: 788–789.

———. 1997. Al-Kh$^w$ārazmī. In *E.I.*$_2$ 4: 1070–1071.

Vicaire, M. H. 1937. Les Porrétains et l'avicennisme avant 1215. *Revue des sciences philosophiques et théologiques* 26: 449–482.

Von Rospatt, Alexander. 1998. Einige Berührungspunkte zwischen der buddhistischen Augenblicklichkeitslehre und der Vorstellung von der Momentanheit der Akzidenzien (*ʿaraḍ, aʿrāḍ*) in der islamischen Scholastik. *Zeitschrift der Deutschen Morgenländischen Gesellschaft*, suppl. 11: 523–530.

Walter, Michael L. 2009. *Buddhism and Empire: The Political and Religious Culture of Early Tibet*. Leiden: Brill.

Walter, Michael L., and Christopher I. Beckwith. 2010. The Dating and Interpretation of the Old Tibetan Inscriptions. *Central Asiatic Journal* 54, no. 2: 291–319.

Warichez, Joseph. 1932. *Les* disputationes *de Simon de Tournai*. Louvain: Spicilegium Sacrum Iovaniense Bureaux.

Watt: *See* Al-Ghazālī, Abū Ḥamīd Muḥammad.

White, Lynn, Jr. 1960. Tibet, India and Malaya as Sources of Western Mediaeval Technology. *American Historical Review* 54: 515–526.

Willemen, Charles. 2006. *The Essence of Scholasticism: Abhidharmahṛdaya. T1550*. Rev. edition with a completely new introduction. Delhi: Motilal Banarsidass.

Willemen, Charles, Bart Dessein, and Collett Cox. 1998. *Sarvāstivāda Buddhist Scholasticism*. Leiden: Brill.

Wink, André. 1990. *Al-Hind: The Making of the Indo-Islamic World*, vol. 1, *Early Medieval India and the Expansion of Islam, 7th–11th Centuries*. Leiden: Brill.

Wippel, J. F. 1985. Quodlibetal Questions, Chiefly in Theology Faculties. In Bernardo Bazàn, John W. Wippel, Gérard Fransen, and Danielle Jacquart, *Les questions disputées et les questions quodlibétiques dans les facultés de théologie, de droit et de médicine*, 153–222. Typologie des sources du moyen âge occidental, fasc. 44–45. Brepols: Turnhout.

Wisnovsky, Robert. 2005. Avicenna and the Avicennian Tradition. In Peter Adamson and Richard C. Taylor, eds., *The Cambridge Companion to Arabic Philosophy*, 92–136. Cambridge: Cambridge University Press.

Wood, Rega. 2010. The Influence of Arabic Aristotelianism on Scholastic Natural Philosophy: Projectile Motion, the Place of the Universe, and Elemental Composition. In Robert Pasnau, ed., *The Cambridge History of Medieval Philosophy*, vol. 1: 247–266. Cambridge: Cambridge University Press.

Wulf, Maurice de. 1956. *An Introduction to Scholastic Philosophy*. Trans. P. Coffey. New York: Dover.

Wüstenfeld, Ferdinand. 1891. *Der imâm el-Schâfi'í, seine Schüler und Anhänger bis zum J. 300 d. H.*, vol. 2. Göttingen: Dieterich.

Yoeli-Tlalim, Ronit. 2010. On Urine Analysis and Tibetan Medicine's Connections with the West. In Sienna Craig, Mingji Cuomu, Frances Garrett, and Mona Schrempf, eds., *Studies of Medical Pluralism in Tibetan History and Society*, 195–211. Halle: International Institute for Tibetan and Buddhist Studies.

Zürcher, E. 2007. *The Buddhist Conquest of China: The Spread and Adaptation of Buddhism in Early Medieval China*. 3rd. edition. Leiden: Brill.

# INDEX